Frequently
Asked
Questions
about the
Universe

Also by Jorge Cham and Daniel Whiteson

We Have No Idea:
A Guide to the Unknown Universe

Frequently Asked Questions about the Universe

Jorge Cham and
Daniel Whiteson

RIVERHEAD BOOKS NEW YORK 2021

RIVERHEAD BOOKS
An imprint of Penguin Random House LLC
penguinrandomhouse.com

Library of Congress Cataloging-in-Publication Data

Names: Cham, Jorge, author. | Whiteson, Daniel, author.
Title: Frequently asked questions about the universe /
Jorge Cham and Daniel Whiteson.
Description: New York : Riverhead Books, 2021. | Includes index.
Identifiers: LCCN 2021001503 (print) | LCCN 2021001504 (ebook) |
ISBN 9780593189313 (hardcover) | ISBN 9780593189337 (ebook)
Subjects: LCSH: Cosmology—Popular works.
Classification: LCC QB982 .C426 2021 (print) |
LCC QB982 (ebook) | DDC 523.1—dc23
LC record available at https://lccn.loc.gov/2021001503
LC ebook record available at https://lccn.loc.gov/2021001504

Printed in the United States of America
1st Printing

Book design by Alexis Farabaugh

For Oliver

J.C.

✦

For Silas and Hazel, whose constant stream of questions
inspired and interrupted the writing of this book

D.W.

Contents

Frequently Asked Questions about the Universe

A Frequently
Asked Introduction

E veryone has questions.

It's an inherent part of being human. As a species, we may not agree on much: politics, favorite sports team, best place to get a taco at twelve a.m. But one thing brings us all together: a need to *know*. We all wonder, and deep down we all have the same questions.

Why can't I travel back in time? Is there another version of me out there? Where did the universe come from? How long are humans going to be around? And who eats tacos at twelve a.m. anyway?

Fortunately, we have answers.

Science has made incredible progress over the last few hundred years, and there's a lot we can say about some very fundamental questions about the universe. There are, of course, still huge mysteries (see our previous book, *We Have No Idea: A Guide to the Unknown Universe*), but for our species, things in the understanding-the-universe department seem to be going in the right direction. So much so that we felt it was time that somebody compiled a list of easy-to-read, cartoon-laden answers to some of humanity's most frequently asked questions.

In this book, we'll explore answers to some of the deepest and most existential questions that people can ask about themselves, the planet, and the nature of reality itself. Have you ever wondered why aliens haven't visited us (assuming they haven't)? Or whether you are truly unique, or just a preprogrammed simulation in some alien video game? Do you stay up at night, wondering if life after death is possible? In your hands are answers to all of these questions.

Each chapter covers a frequently asked question, hopefully revealing in the process some mind-blowing truth about our amazing universe. Think of this book as a primer for your next cocktail party, or as a quick, fascinating read while you sit on the toilet (thankfully, we made each chapter fairly short).

You might wonder what makes us qualified to answer these questions. Rest assured that we have that utmost of qualifications to be authorities on a given topic: we have a podcast.

In our humbly titled, twice-weekly audio program, *Daniel and Jorge Explain the Universe*, we cover topics ranging from microwaves to intergalactic phenomena to hypothetical fundamental particles.

But it's answering questions from listeners that really inspired us to write this book. For us, that's one of the most exciting parts of having a podcast. Nothing brightens our day more than opening our in-box and reading a thoughtful question from a curious listener.

And questions we definitely get! The question askers vary in age (nine to ninety-nine), occupation, and location. You might be surprised at the amazing questions that a nine-year-old from Devonshire can have about the observable universe.

It seems that asking questions and the desire to know is in our hearts. Many would say that wondering about the nature of our cosmos and our place in it is one of the joys of being alive. Of course, it might be frustrating to not know the answers right away, or to only end up with more questions (as in some of the answers in this book), but there's power in just asking the questions.

You see, asking questions supposes that it's possible to find the answers, which we believe is an act of hope. What could be more hopeful than believing that the universe and all its wondrous mysteries can one day be unraveled and understood?

So join us as we plug into the collective curiosity of your fellow humans and take a dive into the questions that frequently stump them. The answers will sometimes be surprising, and they might challenge your view of the universe. Other times the answers will be agonizingly incomplete because they push up against the edge of human knowledge.

In all cases, just remember that most of the fun is in asking the question.

Enjoy!

PS: Don't forget to flush.

Why Can't I Travel Back in Time?

Actually, who said you can't travel back in time?

It's a very common wish to be able to travel back in time. Who among us wouldn't want to go back and talk to famous figures in history, or watch important moments happen in person? You could figure out who really killed JFK or what ended the dinosaurs.

More practically, it would be great to go back in time for smaller stuff, like fixing a mistake you made. If you spilled coffee on your pants, you could go back in time and . . . not spill it. If you said something to your boss you now regret, just go back and not say it. If you ordered a pizza with pineapple on it and then realized that

it's actually gross, you could go back and order a real pizza. It'd be like having an undo button (the equivalent of Ctrl+Z, or Command+Z for Mac snobs) for the universe.

And yet, so far, scientists have not built such a device. The past remains unchangeable. Time is still our great enemy, and it seems we are doomed to live forever in regret about our past mistakes. There are no do-overs in this universe.

But why is that? Why does it seem like we can change the future but not the past? Is there a deep law of physics that makes time travel impossible, or is it just a matter of solving a hard technical problem? And what's the difference anyway?

Well, you might be pleasantly surprised to learn that time travel has *not* actually been ruled out by physicists. It actually *is* technically possible to go back in time. It doesn't work the way you've seen in movies, but it might not be impossible to build a rewind button. In fact, at the end of this chapter we describe a brand-new, physicist-approved idea for time travel.*

So strap on your time machine goggles, prepare your hoverboard and DeLoreans, because we are about to answer that timeless question: Why can't I travel back in time . . . yet?

FAMOUS TIME MACHINES

| H.G. WELLS | THE DELOREAN | TIME TURNER | THE HASHTAG |

---

* Approved by one physicist, at least.

Practical Versus Possible Versus Not Impossible

First, let's clarify what we mean when we ask if something is "possible." It depends on who you ask.

If you ask an *engineer* whether something like time travel is possible, they will say yes if they think they can build a time machine for less than a trillion dollars and do it in less than a decade.

But if you ask a *physicist* whether something is possible, they look at the question differently. A physicist will say something is possible if they don't know of a law of physics that prevents it.

For example:

TASK	ENGINEER	PHYSICIST
COOKING A TURKEY WITH NUCLEAR WEAPONS	HARD, BUT MAYBE	OF COURSE
BAKING A CAKE THE SIZE OF A MOUNTAIN	NO	TOTALLY POSSIBLE
FLYING WITHIN 100km OF THE SURFACE OF THE SUN	PLEASE NO	NO REASON WHY NOT
HOLLOWING OUT THE CENTER OF THE EARTH TO MAKE A GIANT ZERO-G FUN PARK	I QUIT	APPROVED

Since this is a book about physics and the universe, we take the physicist's point of view. That means that our goal in this chapter is to figure out whether time travel breaks any laws of the universe, not whether it would take 14.7 bazillion dollars and hundreds of years to make it real. We trust that once the physicists declare it possible, the engineers will eventually figure out a way to make it practical. Then the next step is for them to hand it off to the software people, who can code an app for it ("Siri, unspill my coffee").

To figure out if time travel can be physicist-approved, first we need to think about time the way physicists do. Time is a very slippery subject, one that has confused and baffled people for a long . . . well, time. Basically, physics thinks about time as the thing that allows the universe to change. It's the *flow*, the motion, the way that *then* turns into *now*. It's what orders and organizes a series of still photos into a smooth movie.

Because the universe does seem to flow smoothly. It doesn't just jump wildly from one moment to a dramatically different moment. You're not on your couch reading this book and then all of a sudden sitting on the beach. That's because the past puts limits on what can happen in the present. If you were sipping coffee a moment ago, then the possibilities for the present include that you enjoy the coffee or spill it on your pants. It doesn't include that you suddenly transform into a blue dragon drinking fermented celery juice.

The past controls the kinds of futures we can have. That's called "cause and effect," and it's at the core of how physics tries to make logical sense of this crazy, bonkers, coffee-stained universe and how it changes.

Those changes happen smoothly and *require time*. Nothing in this universe is instantaneous. Events are connected to one another. When you want to make a pizza, there's a process. You can't just snap your fingers and turn some flour, tomatoes, and cheese instantly into a pizza. The universe requires you to go through the motions: you have to mix the ingredients, knead the dough, cook the tomatoes, drink wine, bake, and so on.* There are steps you have to follow to change from one configuration (raw ingredients) to another (hot pizza). Time is what connects those steps, and without it, the universe just doesn't make sense.

With that understanding of time, let's think about some possibilities for time travel.

You Can't Go Back to the Future

One of the most tempting reasons to want to time-travel is to jump to the past and change something, hoping to influence the future. Like not spilling your coffee, or buying shares in Netflix instead of Blockbuster Video (RIP). You'd like to make a change in the past and then jump back to the present and enjoy the fruits of your manipulations.

There's one big problem with this concept. Quite simply, it just doesn't make sense.

Thinking about time as how the universe flows (or how the pizza bakes), we can see easily that changing the past is nonsense. Let's say you wake up one morning at eight a.m. and you make yourself some coffee. The only problem is that the coffee is bad. So you decide to

---

* Okay, drinking wine is not strictly required by the universe.

TIME: 8:00AM 8:10AM 8:10AM

Hop into
time machine

hop into your time machine, go back to eight a.m. today, and make tea instead of coffee.

This makes sense if you're watching this happen in a movie, but it doesn't make sense from a physics point of view.

From a physics point of view, a configuration of the universe exists (the one in which bad coffee was created) that is not connected to past configurations of the universe. If you made tea instead, how did the bad coffee get made? To a physicist, this breaks the law of cause and effect: there's an effect (bad coffee) but no cause (you made tea instead). In other words, it's like you made a pizza without ever mixing the ingredients.

Unfortunately, this makes changing the past impossible. Breaking the law of cause and effect means that the universe is not consistent with itself, which is a big no-no for physicists.

Now, you might be thinking, *But what about split timelines! Alternate histories! I saw this happen in the Avengers movies!* Unfortunately for Doc Brown (and Iron Man), this also doesn't make sense. How can you change a timeline, or create a new one, when the very idea of change depends on time itself? Timelines *represent* change, so they can't themselves change. And while the concept of a multiverse is something scientists seriously consider, the possibility that we can move or choose between alternate universes is not.

So there are lots of reasons why physics says you can't suddenly jump to another time and change things, which means your dream of manipulating the stock market and getting rich off of physics just went up in a cloud of smoke.*

Where There's a Physicist, There's a Way

Does being strict about cause and effect mean that time travel is *impossible*? Actually, no! It just means that *changing the past* is impossible. What if we wanted to go to the past without changing anything? That might actually work. Let's say you wanted to see the dinosaurs, or skip ahead and see what the future is like. Is that possible? According to our current understanding of physics, that is totally possible (just don't ask the engineers if it's possible yet, though).

To understand how this could work, you have to get used to thinking about space as more than just space. Physicists like to think of space and time together as something called (not very imaginatively) "space-time."

We are used to moving through space near the surface of the Earth, where things are simple. You throw a ball up, it comes down. You walk sideways, you go sideways. Time is equally simple here on Earth: the clock ticks forward and clocks around the world agree with one another.

But physics tells us that in some parts of the universe, space gets really weird. And in those cases, it's best to think of it as being

---

* Getting rich off of physics was never a realistic dream anyway.

joined together with time. To a physicist, we aren't just moving through space *in* time; we are moving through *one* thing called space-time.

And space-time is weird. It does things that are hard for our minds to imagine, like it can *bend*. And fold on itself. It can even loop around.

Let's explore a couple of ways in which this weirdness of space-time could allow time travel.

Infinitely Long Cylinders of Dust

According to Einstein, space-time bends whenever you have something massive around. That's his idea of gravity: it's a distortion of space and time instead of a force. For example, the moon goes around Earth not because our gravity is pulling on it but because it's coasting around a funnel of space-time bent by Earth's mass, like a race car doing laps on a curved track.

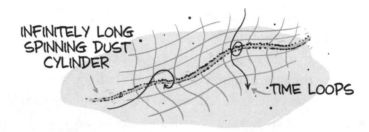

INFINITELY LONG
SPINNING DUST
CYLINDER

TIME LOOPS

But mass doesn't just bend space; it also stretches and squeezes time. And weird configurations of mass can do super-strange things with time. For example, if you make an infinitely long cylinder of spinning dust, you might be able to do something amazing: near that weird column of spinning dust, time and space would bend in a way that lets you move in a loop through time. That means that an object could potentially travel on a path that takes it back to where—and when—it started.

Wormholes

Our modern version of space-time can also be bent and distorted in other weird ways. Space-time can fold onto itself and create a tunnel, or a shortcut, between different points in space-time. This shortcut is known as a "wormhole." You can think of a wormhole as a distortion or a rearrangement of space-time that connects two different points together.

Most people think of wormholes as connecting different points in space (which makes them potentially useful for traveling to far-away galaxies). But in theory, wormholes can also connect different points in time. Remember, it's all one big thing called "space-time." A wormhole could not just take you to your favorite boba tea place across town; it could take you there before boba tea ever became trendy.

HOW TO WORM YOUR WAY INTO
THE FUTURE

No Do-overs?

What's amazing about the two possibilities we mentioned on the previous page is that they allow time travel without breaking the laws of physics. As long as you're not trying to change the past, you could move around in this bent space-time and it could take you to the past (or the future).

The caveat is that it would take you to the same space-time you were in before (you just took a shortcut, or went on a loop), which means that you couldn't change the past *even if you wanted to.* Maybe you did go back and tell your eight a.m. self not to make coffee, but if you did, you would remember that because you are both part of the same timeline. The fact that you made bad coffee, and don't remember meeting your future self warning you about it, means that your future self never went back in time in the first place.

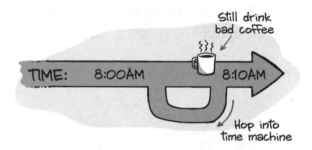

Could we actually do this? The truth is that physicists don't know! This is in the category of "not known to be impossible* but, as far as we know, totally impractical." Nobody has built an in-

---

* We should mention that some physicists think that Einstein's theory is a little bit wrong and that these time loops might be impossible.

finite cylinder of dust. We don't really know how to find worm-holes, much less open one and control it. But the cool part is that "not known to be impossible" means that it's still possible, and while you couldn't unspill your coffee, you could still visit the di-nosaurs or see what the future is like.

Going with the Flow

At this point you might be a little disappointed that while time travel might be possible, it's not the kind of time travel you were probably hoping for. Sure, looking at live dinosaurs might be cool, but how much fun can it be if you have to look at them with coffee spilled all over your pants?

To that end, we are now proud to present to you our brand-new idea for a different kind of time travel, one that might actually al-low you to have an undo button *without* breaking cause and effect. Sure, we came up with this idea just for this book and have spent a grand total of a few hours thinking about it, but hey, all great physics ideas have to start somewhere, and besides, at least one of us is a trained physicist.

Are you ready? Here it is: What if we could *reverse the flow of time*?

You see, physics has a lot of laws that determine how the universe

changes with time. But all of these laws assume the flow of time. None of them actually tell us *how* the flow of time works. For example, we don't know why time flows in one direction (forward) and not the other. In fact, we don't know that time *has* to move forward. Almost all of the laws of physics work perfectly well in both directions.

Almost. There are one or two laws of physics that seem to work differently forward and backward. For example, the second law of thermodynamics says that things tend to get less organized over time and heat spreads out. It's the reason why it's more likely that you break a glass than unbreak it.

But this law doesn't actually require time to go forward. It just says that *if* time were to flow backward, disorganization would have to decrease. That might be weird to see, and we've never seen time flow backward, but physics can't rule it out.

Which brings us to our idea: What if you built a device that could reverse the flow of time selectively? For example, it could reverse the flow of time *inside* of the machine. The machine itself wouldn't travel or go anywhere. To someone sitting outside, the machine would just sit there and still be there later. But *inside* the machine, the rules would be different. Time would flow backward, and particles inside would do the opposite of what they normally would in a time-forward universe.

If you could control the flow of time this way, it would be possible to undo certain things that happen. For example, you could install your office inside the machine and set it to normal time flow. If at any point you spill your coffee, you could tell the machine to reverse the flow of time for a short while. The rest of the universe would continue flowing normally, but inside, your coffee would be unspilled. When the machine switches back to normal flow, you would find yourself with clean pants. Of course, your thoughts would also go backward, so you might want to leave a note to yourself outside the machine to be more careful with your coffee.

It might be hard to tell the difference between traveling back in time and reversing the flow of time in a particular spot, but from a physics perspective, the difference is important. It's not that you or the machine *travels* to a different time (thus breaking cause and effect); you're just reversing the flow in a confined space. If the flow of time is like a big river, this is like creating small eddy currents here and there that temporarily flow backward.

And if this scenario feels a little limiting to you, let's take our imaginary technology to the next level. What if you made a machine that was powerful enough to do the opposite? What if it could reverse the flow of time of the *entire universe*, except for what's inside the machine? Then you could climb inside, hit the button, and watch the entire universe around you flow backward. When you come out, you'd be coming out to a version of the universe that is technically younger (although it de-aged without you).

TIME-TRAVEL SELFIE

What could you do in this younger universe? You could buy Netflix stock, or hang out with JFK, or give up coffee.*

Is this a crazy idea? Yes. Do we know how to make time flow backward, or make entropy decrease? No. Would it work? We have no idea. Is it impossible? Not according to known physics!

That means that, engineers, it's over to you.

---

* Honestly, if you had done this earlier, it would have saved us a lot of trouble.

Why Haven't Aliens Visited Us? *Or Have They?*

A re you excited for aliens to arrive on Earth, or are you terrified?

THE AUTHORS AGREE TO DISAGREE

There's a lot to be excited about if aliens ever visit us. Think about it: if aliens are able to cross the vast distances of interstellar space and find us, that means that they are much more advanced than we are. Imagine the questions we could ask them! How does the universe work? How did it begin? How did you figure out how to travel the stars? Why do some people put pineapple on their pizzas? Wouldn't it be amazing if aliens showed up and just told us the answers? We could skip hundreds or thousands of years of painstaking physics work* and get the answers *now*.

But hold on a moment. What if aliens visiting us doesn't turn out as well as we hoped? A visit from an advanced alien civilization could also be terrifying. Just look at human history. What usually happens when a more advanced civilization meets another civilization? Do they share their knowledge and intellectual wealth and peacefully enjoy a few snacks together? No. It usually doesn't go very well for the civilization being "explored."

In either case, it's bound to be a momentous occasion, which makes us wonder: Why haven't the aliens visited us already? After

---

* Hey, sitting around drinking cofee is hard work.

all, the chances that life exists out there in the universe are pretty large. There's an incredible number of stars just in our galaxy (approximately 250 billion), and there are trillions, if not an *infinite number*, of galaxies out there. And about one out of every five stars has a planet that's just like Earth, which means there are quintillions of (if not infinite!) opportunities for life to develop. The chances that Earth is the only place in the universe where life, even intelligent life, formed seem pretty small.

So then why haven't aliens visited us? Are they avoiding us, or is the universe just too big for neighborly visits? And how would they even find us?

To figure this out, let's look at four possible scenarios.

Scenario #1: They Heard Us and They're Coming to Find Us

One possibility is that aliens have heard us and they're already on their way. Maybe the aliens are good listeners and they've picked up some of the radio and television transmissions we've been inadvertently broadcasting into space. Intrigued and charmed by our

Nanu nanu! Mork... Ha-ha! Those humans!

humor and culture, they immediately launch a ship and are heading straight for us as we speak.

What does physics have to say about this scenario? Is it possible for aliens to have detected our signals? And has enough time passed for them to have made it here by now?

One limitation is that we haven't been broadcasting radio signals for very long. Our species started broadcasting radio and television and other signals approximately a century ago. And while the speed of light seems very fast to you while you are stuck in traffic fantasizing about getting home, space is very big. So even light-speed messages will take a long time to reach any potential alien worlds.

And even if they've heard our messages, it would take them a long time to come visit us.

Let's think about the physics of their trip. We'll start by assuming that they have some sort of spaceship that travels at a decent fraction of the speed of light (let's say half, or about 150 million meters per second). You might worry about the time it takes for them to accelerate to that enormous speed, but, somewhat surprisingly, this is a small portion of their trip. Even if they are squishy beings like us who can't withstand acceleration forces greater than a few times Earth's gravity without turning into pudding, they could still spend the majority of their trip going at top speed. For example, you can get up to half the speed of light in less than a year accelerating at a modest 2-g (twice the acceleration due to Earth's gravity).

Now let's do the math. Since we've only been transmitting radio signals for about 100 years, any aliens arriving anytime soon would have to live within about 33 light-years of us: it would take 33 years for our signal to get there at the speed of light, and approximately 66 years for them to travel here on their spaceship (which we assume can go at half the speed of light). In this scenario, any aliens living *farther* than 33 light-years from us haven't had a chance to get here, because not enough time has passed for them to get the message *and* make the trip.

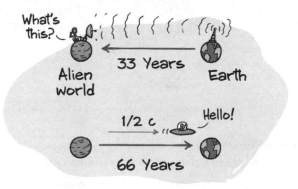

Could there be aliens living within 33 light-years of us?

We know that the closest star system to us (Proxima Centauri) is just over 4 light-years away. And it just so happens to have an Earth-size planet orbiting around one of its stars. If there *are* aliens there that have heard our signal, they would have had plenty of time to hop on a spaceship and come visit us. So why haven't they? One theory: they were waiting for the series finale of *Lost*, which aired in 2010 and arrived at their planet in 2014. That means we can expect them to get here to complain about it in 2022.

What if we look farther out? Within 33 light-years, we know there are just over 300 star systems, approximately 20 percent of which are likely to have an Earthlike planet (defined as a planet that's about the same size and cozy distance from its star as ours). That means there are around 65 Earthlike planets that could have heard our earliest radio signals and sent a delegation of aliens to us by now.

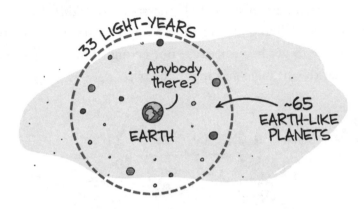

But they haven't. Why not?

Of course, there are many reasons why aliens could have heard our signal but haven't come to visit us. Maybe they didn't like what they heard, or maybe they're just not interested, or can't be bothered. But it's hard to imagine that an intelligent civilization, probably as isolated as our own, wouldn't jump at the chance to reach out to, or at least check out or respond to, one of its neighbors.

The fact that we haven't been visited by an intelligent alien civilization responding to our radio signals suggests perhaps a more obvious truth: that there *aren't* any intelligent alien civilizations within such a close distance to us. It tells us that maybe the odds of finding advanced intelligent life is less than 2 (us and another civilization) in 65 planets. This seems like the most likely explanation. After all, looking back at the history of life on Earth, and the precariousness of our civilization, the odds of us even being here seem much smaller than 1 in 32.5.

Scenario #2: They Stumble upon Us

If the reason we haven't been visited by aliens is that there aren't any within listening distance of us, then maybe we need to think of another reason or another way in which aliens could find us. After all, our radio signals have only reached a tiny little bubble of the galaxy: about 100 light-years in all directions, whereas the Milky Way is over 100,000 light-years wide. It's no surprise that most of the galaxy has no idea we are even here.

What other reason could an alien civilization, one that lives far beyond the reach of our radio signals, have for coming to visit us?

Well, the galaxy is billions of years old. What if there's an extremely advanced alien species out there that has a fondness for exploring? If they have been exploring for thousands or millions of years, what are the chances that they will stumble upon us?

It's a little hard to imagine why an alien species would spend so much time exploring the galaxy. Maybe they're on the lookout for good television shows, or maybe they're looking for tasty new snacks (hopefully not us), or raw materials, or new places to settle in. Who can guess at the motivations of a billion-year-old alien civilization? But whatever the reason, let's assume they're out there and they're looking.

Could they find us?

Let's make some assumptions about their exploration plan. First, we'll assume they're going to use spaceships. How many ships do they need to send out and how many years would it take for them to visit every planet in the galaxy?

We know that, on average, there is about one Earthlike planet for every 1,250 cubic light-years of space, and that the average distance between these planets is roughly 11 light-years. Sometimes you can find two of these planets in the same solar system; other times you might have to go 50 or 100 light-years between them. For a long voyage, what matters is the average, and that average is about 11 light-years.

Now, if each of their exploring ships travels at half the speed of light, it would take each ship twenty-two years to go from one planet to the next. That means that if you only launch one ship to explore the whole galaxy, it would take you about a trillion years to visit every Earthlike planet in the galaxy. If this is a mission to find tasty snacks, there's no way you're getting home before the snack gets cold.

The good news is that you can easily speed this up by launching more ships. As long as the ships start off in different directions and don't overlap, the more ships you launch, the more planets you can explore.

If you launch 1,000 ships (presumably from a somewhat central location), you can visit every Earthlike planet in the galaxy in about a billion years. As you launch more ships, the time it takes to explore the galaxy keeps falling. If you launched a million ships, it would take you a million years. If you launched a *billion* ships, that number drops to about 50,000 years. After about a billion, launching more ships doesn't help you because it still takes the ships the same amount of time (about 50,000 years) to reach the edges of the galaxy.

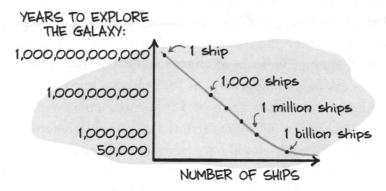

YEARS TO EXPLORE
THE GALAXY:

1,000,000,000,000 — 1 ship

1,000,000,000 — 1,000 ships

1 million ships

1,000,000 — 1 billion ships
50,000

NUMBER OF SHIPS

THE GRAPH THAT LAUNCHED A BILLION SPACESHIPS

Fifty thousand years may sound like a long time, but compared to the age of the galaxy (13.5 billion years) and the age of our planet (4.5 billion years), that's almost no time at all.

This means that if there *is* an alien civilization out there that (a) is actively visiting other planets and (b) has the resources to build a large fleet of exploring ships, then the probability that they would reach us is pretty high. In fact, it means they would probably come around to visit Earth pretty frequently if they are persistent in their search for the perfect snack. Once the ships are spread out through the galaxy, they can visit every planet in less than 50,000 years.

And that's if we assume only *one* such advanced civilization. What if there are a lot of advanced civilizations out there exploring? Then

HEY, WE GOT
HERE FIRST!

the chances that some alien, any alien, stumbles upon us get even higher.

So what does it mean that we haven't been visited yet by an exploring alien ship? We've been smart enough as a species to understand what's happening around us for at least a few 10,000 years (recorded history goes back about 5,000 years, and cave paintings date back more than 44,000 years). If such an exploring ship came to visit, you probably would have heard about it by now.

The fact that we haven't been visited (as far as we know) by exploring aliens tells us that maybe there are no galaxy-exploring civilizations out there. Maybe the reason we haven't been visited by aliens has more to do with economics than physics or biology: maybe space is too big, and the stars are too far away, so that it doesn't make a lot of sense to visit and explore other planets in the galaxy.

Scenario #3: The Aliens Are Very, Very Clever

Okay, so maybe building a massive fleet of a billion ships is too much for any alien civilization. Let's face it, building and manning a billion ships is a lot of work just to find new kinds of snacks. How *else* might aliens ever find us?

Well, there is one other possible scenario, but it requires some extra-imaginative thinking. What if the aliens *are* super smart? And what if they are *so* smart, they've come up with more efficient ways to explore the galaxy?

Hear us out: What if the aliens built *self-replicating* exploring ships?

Imagine ships that flew out into space and then made more of themselves. You could start off with a handful of these self-flying ships and send them out in the direction of nearby solar systems. When they arrive, their first job would be to search the solar system for life. To avoid the hassle of takeoff and landing, you could give them powerful cameras that let them take pictures of the surface of planets from space.

Next, they would search for the raw materials needed to build more of themselves. Our solar system, for example, has plenty of metals and the ingredients for rocket fuel floating in the asteroid belt: huge chunks of iron, gold, platinum, and ice. An AI-controlled ship could gather the raw materials it needs to build and fuel several copies (say five) of itself. Then those five new ships could launch in new directions, repeating the cycle.

Go forth and multiply

This strategy makes the number of ships grow *exponentially*. If you start with five ships, then those five ships become twenty five. After the fifth round, you would have 3,125 ships. After the ninth round, you would have nearly two million ships, and after only thirteen rounds, the number of ships is over one billion.

That means that a single smart alien civilization could send probes that can explore the entire galaxy in less than a million years, and all they have to do is build five starter ships. Suddenly, the economics of this just got better.

Of course, this sounds like pretty complicated technology, but it's something that even human engineers are considering. We are nowhere close to being able to do it, but it may be possible for an older and more advanced civilization. Who knows, in a few hundred years, even *we* might be able to build such ships.

The important thing is that it only takes one civilization to start the cascade that leads to a billion ships. That means that if there *are* aliens out there, and if they are clever enough, then the chances that we've been visited by one of their self-replicating ships is fairly high.

Of course, the fact that we haven't seen any such probe arrive and announce itself could mean any number of things. Maybe no such advanced civilization exists out there and it's up to us to develop this idea. Or maybe there *are* advanced civilizations out there, but they thought that this was a terrible idea.

After all, maybe they don't *want* us to know they exist . . .

Scenario #4: Or Have They?

In all of the scenarios above, we made one small assumption: that when the aliens come, they will announce themselves with great fanfare, launching a new era of interspecies harmony (or interspecies conquering).

But what if nearby aliens or alien explorers or self-replicating probes *have* already visited Earth, but we didn't notice? Maybe they came to Earth too early. Life on our planet has been around for billions of years, but there's only been intelligent life capable of recognizing and recording the enormity of an alien visit for a few ten thousand years. What if we missed them? What if they came to visit and we were still in our diapers as a civilization?

If that's true, there's no need to feel like we missed out. After all, there's good reason to think they'll come back. They will almost certainly have noticed life brewing on their first visit to Earth, since life started soon after the planet formed, which gives them a reason to check back on us. Remember that a large fleet of ships can explore the galaxy every fifty thousand years, so maybe we just have to wait a little more for the bus to come around again.

But hold on. What if the reason we didn't notice their visit is that

they don't *want* us to know they visited? What if they don't want to talk to us? What if our basic assumption is wrong and they are not looking for someone to hang out with? The physics of galactic exploration can't rule out sneaky, or shy, alien visitors. Maybe they know better than to go around mingling with potentially dangerous alien species. (Yes, what if *we're* the bad-news aliens?) We can't expect to understand how aliens may think.

To summarize, there are lots of reasons why aliens haven't visited us (or why they haven't told us they visited us). It's a pretty big galaxy, and an even bigger universe, and there's a lot we don't know about the chances that there is intelligent life out there. There is still the possibility that we are the smartest species in the galaxy (or the universe) and that other aliens aren't likely to come visit us anytime soon.

If that's the case, maybe it's up to us to go out there and visit other aliens. If not for the sheer joy of exploration, let's at least do it for the snacks.

Is There Another You?

Wouldn't it be strange if there was another copy of you somewhere out there?

Someone else just like you, with the same likes (bananas) and dislikes (peaches), the same skills (amazing banana smoothies) and flaws (won't shut up about banana smoothies), the same memories and sense of humor and personality. Would it be weird to know that they exist? Would you want to meet them?

Or, stranger still, imagine there was someone out there who was almost exactly like you, but a *little* bit different. What if there's a version of you that is better? Maybe they make tastier fruit smoothies or lead their lives in a more meaningful way. Or what if there's a version of you that is less talented—or meaner, like an evil twin?

Is that possible?

It might be hard to imagine, but the possibility that there's another you is something that physicists can't rule out. In fact, physicists don't just think another you might be possible; some think it's *more likely than not.* That would mean that right now, as you read this, there could be another you somewhere out there, wearing the same clothes you're wearing, sitting in the same way you're sitting, and reading the same book you're reading (well, maybe a slightly funnier version).

To get a sense of what this means and how likely it is, let's start by considering just how special you are.

The Odds of You

At first, it may seem pretty unlikely that someone exactly like you could exist out there. After all, imagine all the things that had to happen just for the universe to make you.

A supernova had to explode near a cloud of gas and dust, shocking it into gravitational collapse to form our Sun and our solar system. A tiny little clump of that dust (less than 0.01 percent) had to gather together to form a planet at just the right distance from the Sun so that water wouldn't freeze or turn into steam. Life had to begin, dinosaurs had to get wiped out, humans had to evolve, the Roman Empire had to collapse, and your ancestors had to dodge the black plague. Then your parents had to meet and actually like each other. Your mother had to ovulate at just the right time, and the sperm with the other half of your genes had to win a sprint against billions of others. And that's just getting to your birth!

Think about all of the decisions you've made in your life that have led you to be who you are today. You ate a lot of bananas, or didn't. You met that one important friend, or didn't. You decided to stay home that time when you otherwise would have been run over by a fruit cart. You somehow found this silly book about the universe and decided to read it. All of it, starting 4.5 billion years ago, led to your existence right here, right now.

What are the chances that all of those things would happen in the *exact* same way *again* to make another you? It seems pretty unlikely, right?

Maybe not! Let's trace back through all of the random events and decisions and moments that led to you, and try to calculate what those odds are.

We'll start with today: How many decisions have you made since you woke up? You probably decided to get out of bed, picked what to wear, and chose something to eat for breakfast. All of those decisions, even if they seem small, can change the course of your life. For example, your choice to wear that blouse or tie with the banana pattern on it might mean the difference between your future spouse noticing you or not. Let's assume that you make approximately one or two potentially life-changing decisions every minute. It sounds stressful, but if you subscribe to quantum physics and chaos theory, that number should be a *lot* higher. Let's assume it's only a couple per minute, which means that you make thousands of such important decisions a day, up to about a million per year. If you are more than twenty years old, you have made over *20 million decisions in your life* to get where you are today.

It's a life-changing decision!

Next, let's assume that each decision you made had only two possible outcomes, like A or B, or bananas and peaches. Really, it's many more (have you seen how many choices a typical brunch menu has these days?), but let's keep it simple. To calculate the odds of you being who you are because of those 20 million decisions, you have to raise 2 to the power of 20 million, or $2^{20,000,000}$.

Why? Because the number of possibilities grows with each decision. If you have to choose which side of the bed to get off on (right or left), what fruit to have for breakfast (bananas or peaches), and how to get to work (train or the bus), then there are 2 x 2 x 2 (or 2^3) ways in which your day can go. The odds that you got off on the left side, had bananas, and took the bus is 1 in 2^3, or 1 in 8.

So if you make 20 million "A or B" decisions in your life, that means that there are $2^{20,000,000}$ different ways in which your life could have turned out. That's a really big number. And we're just getting started!

We also have to include the odds of you being born, which is the result of the decisions your parents made. If we include them, you have to add another 40 million decisions (20 million for each parent). Bring in your 4 grandparents, and you have 80 million more. Great-grandparents? Another 160 million. See where this is going? The number of ancestors doubles per generation, adding even more

SO I CAN BLAME MY PARENTS FOR EVERYTHING?

decisions that potentially influenced you. Humans have been on the planet for at least 30,000 years, or roughly 1,500 generations. Factoring in all of *their* decisions is going to make our number even bigger.

Actually, it gets a bit complicated to count, because if you go far enough back, some of your relatives are related to other relatives, which means the same person can appear in your family tree twice. That's an awkward topic to talk about, and it also makes the math harder. For simplicity, we'll assume you're only affected by 2 people per generation. That's still 1,500 generations × 2 people × 20 million decisions = 60 billion decisions. Now the odds of you happening are 1 in $2^{60,000,000,000}$.

But why stop there? Let's take into account prehuman history and the billions of years of evolution going back to the smallest microbes. Life on Earth began about 3.5 billion years ago. If you had to make a family tree that far back, it would be mostly microbes and simple plants. They may not have made conscious decisions, but they were affected by random events: the way the wind blows, whether the sun shines on them, if rain falls, and so on. Let's assume that at a minimum, your microbial ancestors were affected by one random event per day and that each of those random events also had two possible outcomes (e.g., a rock falls on your microbial ancestor or not). That means we have to add another *trillion* (1,000,000,000,000) decision events to our odds.

Now let's unwind our local blob of the universe all the way through the formation of the solar system, 4.5 billion years ago. Keep going to the previous stars or planets that held the atoms that make up you, then go all the way back to the Big Bang, 14 billion years ago. As a dramatic underestimate, let's assume again that on each of those days, one important event happened that could influence the direction of your life. That means that there are roughly a *quadrillion* decision events leading up to today, which means the odds of you being here just jumped to about 1 in $2^{1,000,000,000,000,000}$.

Unlikely but Not Impossible?

That number, $2^{1,000,000,000,000,000}$, is a very big number. Try to picture a 1 with about 100 trillion zeros after it. It's so big, our brains can't even comprehend it. For comparison, there are 2^{265} particles in the entire observable universe. To get $2^{1,000,000,000,000,000}$ particles, you'd have to square the entire observable universe about 3 billion times.

When your mother said you were a small miracle, she wasn't kidding! The chance that someone exactly like you has ever existed, or will ever exist again, is 1 in $2^{1,000,000,000,000,000}$, otherwise known as pretty much zero. For you to happen again would be like rolling a die that had $2^{1,000,000,000,000,000}$ sides to it and luckily getting the same number twice. Those are not odds you typically want to bet your house on.

How is it possible then that physicists could think that another you might exist? Well, we live in a strange reality, and there are actually several different ways in which another you could be out there. That includes one scenario where you could actually *meet* them (cue evil twin music: *dun-dun-duuuunnnn*) . . .

The Multiverse(s)

If it's hard to imagine that another you could exist in this universe, then maybe we have to look elsewhere for the peach-loving, train-taking version of you.

A lot of physicists are attracted to the idea that there might be more to reality than just our universe. Maybe, they say, there are actually *multiple* universes. Is it possible that in one of these other universes you can find another version of you? This concept is called the "multiverse," and, ironically, physicists have come up with several different versions of it.

The Different Universes Multiverse

In one version of the multiverse, our universe is one of an infinite number of universes. The only catch is that each universe is a little bit different.

You see, if you look closely at our universe, you'll find that a lot of things about it seem arbitrary and kind of weird. For example, the cosmological constant that controls how the universe expands happens to be 10^{-122}. Why is it that exact value and not something different? As far as we know, its value *could* be different, but there's no apparent reason it isn't, which makes physicists really uncomfortable. To a physicist, every cause should have an effect, so to think that the cosmological constant is 10^{-122} *just because* drives them bonkers.

The only way this makes sense, physicists argue, is if there are other universes where this value is different. For example, maybe there's a universe out there where the cosmological constant is 1. And another universe where it's 42. Each universe gets a random value, and we just happened to get a weird one. That way, the fact that our cosmological constant is 10^{-122} is not that strange. We're just one random sample from a whole infinity of universes.

NOT ALL UNIVERSES WERE CREATED EQUAL

Could one of those other universes have a version of you? This is hard to say.

If you change one of the basic parameters even the slightest amount, how different would that universe turn out? Is it even possible for life to develop in the same way in that universe? It seems possible to have another universe where the difference is so small (say, the cosmological constant is only $1 \times 10^{-1,000,000,000,000,000}$ percent different than ours) that a version of you could have been born. But that begs a different question: Would that version of you be exactly like you if it lives in a universe that's fundamentally different?

The Quantum Multiverse

Another version of the multiple universes hypothesis is the quantum multiverse. This version comes from an attempt to explain another strange thing about our universe: the bizarre randomness of quantum mechanics.

According to quantum mechanics, every particle has an inherent uncertainty about it. For example, if you shoot an electron at another particle, it's impossible to know ahead of time if it's going to bounce left or right. The only way to find out is to actually shoot the electron and then measure which way it goes.

But what makes one electron go left instead of right? Or right instead of left? Once again, we're left with a situation that drives physicists bonkers: an effect without a cause. Does the electron choose which way to go *just because*? Do all particles, when they interact with other particles, choose what to do *just because*?

"Just because" may work at the preschool playground, but to a physicist pondering the universe, this just isn't good enough. Enter the quantum multiverse.

What if, when an electron has to choose to bounce left or right, the universe splits in two? In one universe the electron goes left, and in the other the electron goes right. And what if, in both of those universes, the next time another particle has an interaction those universes split again, making even *more* universes? Believe it or not, this makes more sense to physicists because it means that the universe isn't random. Why did the electron go left? Because there's another universe in which it went right. It's not random, because the electron goes both ways.

What does this mean for our search for another you? If the quantum multiverse is real, it means that there is *most definitely* another version of you out there. In fact, if new universes are being created every time a particle makes a "left or right" decision, more yous are being popped into existence all the time. In the quantum multiverse, there isn't just one you out there; there are countless yous, and more are being created as we speak.

IT'S A QUANTUM BONANZA

Of course, some of those universes could have been created a long time ago, perhaps as long ago as the Big Bang, and they could be so different from ours that a version of you doesn't exist in them. Maybe an electron going left instead of right in the early universe was so consequential that it made a whole branch of the multiverse unrecognizable to ours. Or there could be a branch of the multiverse where a quantum effect somehow steered your life in a totally different direction, in which case there *could* be evil twin versions of you out there, making peach smoothies instead of the obviously better banana version.

Is the Multiverse Real?

In both of those versions of the multiverse, there might be another you. In fact, there might be *lots* of other yous in these other universes. But do we know if these theories are true? Unfortunately, no. So far, the multiverse is just an idea thought up to explain, or at least excuse, why the universe seems to be so choosy. And even if these other universes exist, we aren't connected to them and we don't have any way to interact with them. That means we may never confirm they exist, much less visit them.

So is the long-awaited, soap-opera-worthy, dramatic meetup between you and your evil twin doomed to never happen?

Not necessarily. There is one other way for another you to exist: they might exist in *this* universe. That means it might still be possible for you to meet them (again: *dun-dun-duuuuunnnn*).

Another You in Our Universe

Could another version of you exist in this universe? You know, the same universe you're sitting in? Could it be that right now, as you read this, you are sharing the same space, or even the same galaxy, as your evil twin?

What if, in another part of our universe, there was a cloud of gas and dust that was just like the cloud of gas and dust we came from? And what if a supernova happened in just the right way to make a sun and a solar system exactly like ours? And what if in that solar system, a planet formed just like Earth at exactly the same distance from its sun as our Earth? And what if on that Earth, the exact same things happened that happened in our planet, so that an exact copy of you was made?

Before, we estimated the odds of this happening to be pretty astronomical. We compared it to rolling a die that had $2^{1,000,000,000,000,000}$ sides to it and expecting to get the same number twice.*

Now, those are tough odds for sure, but there is one important thing they are not: zero. That means that, as unlikely and miraculous as you are, it is technically not impossible that you could happen again in this universe. Just because it's hard to roll the same number twice with a giant $2^{1,000,000,000,000,000}$-sided die doesn't mean that it can't or won't happen. Each time you have a cloud of gas and dust that forms into a star is a roll of the die that could potentially

---

* By the way, we did the math: a $2^{1,000,000,000,000,000}$-sided die where each side is 1 cm² in size would be bigger than the entire observable universe.

make another you. In theory, it could happen a few solar systems over or maybe just on the other side of the galaxy. The point is, it's possible.

And the odds of you happening again get even better if we consider more of the universe. For example, there are about 250 billion stars in our galaxy, and that means there are 250 billion more chances for the universe to roll the die and make you again. Of course, rolling the $2^{1,000,000,000,000,000}$-sided die 250 billion times and hoping to hit the same number again would still give you pretty slim chances, but there's lots more universe out there.

Let's consider the observable universe. We know there are at least 2 trillion galaxies in the part of the universe we can see, each with a few hundred billion stars. Now the odds are a little better: now we're rolling the die 2^{78} times hoping to hit 1 in $2^{1,000,000,000,000,000}$.

But what if the universe is a *lot* bigger than what we can see? What if it's so big, and so full of stars, that there are $2^{1,000,000,000,000,000}$ stars in it? That means that you're rolling the $2^{1,000,000,000,000,000}$-sided die $2^{1,000,000,000,000,000}$ times, which gives you pretty good odds. In fact, it's *more likely than not.** If you had a penchant for gambling, you might now consider betting your house on it.

Is the universe that big? Is it possible that the universe has $2^{1,000,000,000,000,000}$ stars in it? Actually, physicists think the universe might be bigger than that. In fact, they think the universe is most likely infinite.

---

* For example, the odds of rolling any given number (say a six) by rolling a six-sided die six times is about 66 percent, which is devilishly peculiar.

An Infinite Universe

An infinite universe is a hard thing to wrap your head around (literally and figuratively). Imagine a universe that goes on *forever*, in all directions.

What does that mean for the possibility of another you? If the universe is infinite, there is most definitely another you out there. It might be good odds to roll a $2^{1,000,000,000,000,000}$-sided die $2^{1,000,000,000,000,000}$ times hoping to hit your number, but if you get infinite rolls, you will most certainly get it. Infinity is a number so big that even numbers like $2^{1,000,000,000,000,000}$ pale in comparison. In fact, if you roll the die an infinite number of times, you will hit 1 in $2^{1,000,000,000,000,000}$ not just once but an *infinite* number of times. That means there wouldn't just be another you out there in this universe; there would be an *infinite* number of yous in the universe.

Imagine hopping on a rocket ship and flying off in one direction. At first, all the stars and galaxies will look very different from one another. That makes sense, because the chances that those stars would form again are pretty small. But eventually, if you look at enough places, even very unlikely things will happen again. You'll find a place that happens to have the same conditions that made our Sun and our planet, and even you. And if you keep going farther, you'll see it again. And again. And again, to infinity. And each time you pass those repeating stars, you'll be able to see other versions of you: versions that are exactly the same, and also different versions. That's how big infinity is.

And all of those yous would be in the same universe, sitting in the same space. Of course, they might be so far away that you could never actually reach them on a spaceship. But what if you could

THE UNIVERSE RUNS OUT OF IDEAS

find another way to cut distances in space? In theory, something like a wormhole, which connects different points in space-time, could take you closer to those other versions of you. This is something physics can't rule out!

GUESS WHO'S COMING TO DINNER?

Wrapping It Up

Is there another you out there? It depends. If the multiverse is real, or if the universe is infinite, then most certainly the answer is yes. If neither theory turns out to be true, then almost certainly the answer is no. What's interesting is that there doesn't seem to be a lot of middle ground. Either there is most likely only one of you in the entire universe or there is an infinite number of you.

Now *that's* a soap-opera-worthy cliffhanger. *Dun-dun-duuu unnnn!*

How Long Will Humanity Survive?

L et's start with the bad news. We're all going to die.

If you were hoping that humanity would last forever, that our civilization and culture would continue and thrive until the end of time, we are sad to tell you that this is highly unlikely.

Sure, humans have come a long way in a pretty short period of time. It seems like just yesterday that we descended from the trees, built cities, invented computers, discovered Nutella, and understood deep truths about the universe. Compared to the age of the universe (13.7 billion years), we only just got here. But how long can this crazy party go on?

Will we live out our days in the golden years of the universe, billions or even trillions of years from now? Or will we snuff out like a rock star in a blaze of Nutella-filled glory?

You see, there is no shortage of things that threaten to end our existence. The universe is filled with perils that could spell doom for us, from self-inflicted annihilation to planet-killing asteroids to getting engulfed by our own sun. To survive as a species until

the end of time means that we don't just have to survive *one* of these things; we have to survive *all* of them.

The good news is that we still have a chance. And that chance depends on two things: the probability that these humanity-ending events can happen, and the timescale that we're talking about. Because while we may dodge the bullet on things that can kill us *right now*, there could be bullets coming at us from deep in space, or even from the fabric of reality itself.

PHEW! DODGED THAT BULLET.

So tear up your outdated Mayan calendars, because we're going all the way on this topic: all the way to the *end*.

Immediate Threats

It might be comforting to imagine humanity living out its days in the waning years of the universe, sitting around eating Nutella sandwiches* billions of years from now, but these days it seems like the world could end at any moment. Open up your web browser on any morning and it feels like catastrophe is right around the corner: global pandemics, crazy dictators, or everyone simultaneously slipping in the shower.

But as catastrophic as these things sound, would they really end the human race? After all, we've survived pandemics before. Dictators don't live forever. The World Health Organization could rally and buy every man, woman, and child a bath mat.

Let's consider the things that, from a physics point of view, could actually end the human race. What are the most immediate threats to human existence *right now*? For our money, they are:

---

* Or Nutella tacos. Discuss.

Nuclear War

Remember back in the 1980s when everyone worried about nuclear weapons? Guess what? They're still here! We might all be distracted by our Twitter or TikTok feeds, but let's not forget we are still one push of a red button away from the end of human civilization. And that's because nuclear bombs are *powerful*. The first ones developed could release sixty terajoules of energy. Since then, they've gotten thousands of times more powerful and we have *a lot* more of them.

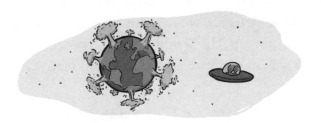

What are the chances of an all-out nuclear war? Higher than you might imagine. There have been many times in history when American or Russian leaders were on the verge of launching a nuclear war. These include the following terrifying incidents:

✦ In 1956, a flock of swans was misinterpreted as a group of Russian fighter jets and, coinciding with several other innocuous events, nearly convinced US officials to launch a counterattack.

✦ In 1962, a Soviet submarine off the coast of Cuba was given warning shots by a US fleet. Thinking it was the beginning of an attack, the sub almost launched its nuclear weapon at the United States.

◆ In 1979, a training program was accidentally loaded into the main NORAD computers, which then transmitted a message to the US president that 250 Soviet missiles had launched and that a decision to counterstrike was needed within three to seven minutes.

◆ In 2003, an elderly woman in the suburbs of London accidentally hacked into US computers while trying to do her grocery shopping, and nearly started a nuclear strike when she entered the ingredients for a triple cherry bomb.

It sounds ridiculous, but all of these things actually happened. Okay, one of these didn't happen, and if you can't tell the difference, then we've made our point. Like something out of a Douglas Adams novel, humanity could very easily have come to an end because of something as silly as a flock of swans. And this is not even the full list of close calls.

How bad would nuclear war be? Very bad. The problem isn't just the explosions and the radiation. The amount of smoke and dust blown into the skies would block the sun and lead to a nuclear winter. Temperatures would fall by tens of degrees for decades, leading to a new ice age on top of the widespread radiation poisoning. Or, if one of the bombs exploded near a body of water, it could send a lot of vapor into the upper atmosphere. This would create a superpowerful layer of greenhouse gases, causing runaway heating and putting Earth on a trajectory to get very hot. Either way, Earth would be uninhabitable for humans.

Climate Change

Even if we somehow avoid blowing ourselves into smithereens, we still have to deal with the impact of our carbon emissions. Climate change is real, and it is man-made. Getting scientists to agree on *anything* is pretty hard, so the fact that ninety-eight out of one hundred scientists believe climate change is happening means that the data must be pretty solid.

Some might shrug off climate change as no big deal. After all, if Earth warms up a few extra degrees, what's so bad about that? Well, if you have any doubts about how serious climate change can be, just ask any Venusian what they think of the subject. What, you don't know anyone alive from Venus? Exactly.

Venus has one of the most inhospitable environments in the solar system. The temperature on the surface of Venus is more than 800°F (462°C), which is hot enough to melt lead. The surprising thing is that scientists think that Venus might once have been a lot like Earth. The two planets were likely made out of the same materials in the solar system, and so Venus might have also had liquid oceans of water and reasonable temperatures. But at some point, maybe due to its proximity to the Sun, the oceans could have evaporated, triggering a runaway greenhouse effect: the water vapor trapped more of the sun's rays, making the planet hotter, which caused more water to evaporate, which made the planet hotter, and so on.

Something very similar can happen here on Earth if we're not careful.

Runaway Technology (i.e., "Whoops")

Let's suppose humans manage to get smart and avoid blowing ourselves up or ruining our planet. Is it possible we could get too smart? Could we somehow invent a technology that actually ends up killing us? As our technology gets more powerful and more sophisticated, some scientists think that's a real danger. We might create an artificial intelligence that decides that we are obsolete and need to be retired. Or we may create the gray goo—a swarm of self-replicating nanobots that gets out of control and eats up all of the organic matter on Earth.* Who knows what other technology we might invent in the near future that could accidentally wipe us out?

* This is a real thing. Just Google it.

Less Immediate Threats

Okay, let's be optimistic and imagine that humans have somehow managed to get rid of nuclear weapons and avoid environmental collapse, and are smart enough to put an off switch in every advanced technology we create. Maybe we become an older, wiser civilization that has put those dangerous devices aside and learned to work together for our common survival. Let's hope so, because pretty soon something else will be coming for us.

If we survive the dangers here on Earth, then on the timescale of thousands of years other threats start to become more real. Namely, death from space.

What if a big asteroid emerged from the depths of space, struck Earth, and caused massive destruction? It's happened before (remember the dinosaurs?) and it could happen again. It could be a really massive rock, big enough to crack Earth itself. Or it could be a modest, Manhattan-size asteroid that kicks enough dirt into the atmosphere to cause radical environmental change. As we'll cover in a later chapter (Is an Asteroid Going to Hit Earth?), this is not something that we foresee happening in the next few hundred years (we are currently tracking most Earth-killer-size asteroids in our solar system), but over the next thousand years, who knows? The predictions get very fuzzy deeper into the future.

More alarming, something else might be coming for us. Comets have orbits so large that there are many in our solar system we still don't even know about. One of these comets could hit us when it comes back around in its thousand-year orbit.

In either case, let's hope that Bruce Willis is still around then,[*] because we'll need some way to either deflect or destroy those asteroids or comets if we hope to survive the next few thousand years.

Million-Year Threats

What about on the scale of millions of years? What are some of the threats that become more likely if we somehow manage to survive that long?

Well, the universe is a dangerous place, and there are things out there that could wipe us out even if we somehow learn to clone Mr. Willis and have *Armageddon*-style asteroid and comet preparedness plans. One real danger is that our entire solar system could get scrambled by a visiting object from deep space.

You see, the planets in our solar system are in nice, cozy orbits

---

[*] Have you noticed the guy doesn't seem to age?

around the Sun, and those orbits are important. They're also very fragile. Imagine that each planet's orbit is like a plate spinning on the tip of a finger. That means that the solar system has eight of these plates spinning at the same time. What happens if a big and heavy visitor comes and bangs things around? It could be a solar system–size disaster.

A small visitor, like the interstellar comet 'Oumuamua, is not going to cause any real disruption. But suppose a really, really big asteroid (maybe a rogue planet from far away) were to enter our solar system.

The bad news is that a rogue planet like that doesn't even have to hit anything to kill us. It could disrupt the solar system simply by getting too close. Its gravitational pull might be enough to throw other planets' orbits off, causing chaos and disorder in our quiet little neighborhood.

In fact, it wouldn't take much to make things bad for us. Earth's orbit around the Sun is so fragile that a small tug from an unex-

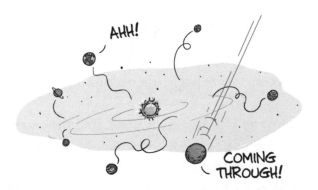

pected visitor might be enough to change it. We could end up too close to the sun (frying everything on Earth) or too far away (freezing everything). More drastically, if it comes close enough, it could end up ejecting us from the solar system, leaving us to float out in space forever.

At the scale of millions of years, we can let our imaginations fly even more. What if instead of an asteroid coming to disrupt our solar system, it's another star? Or even a *black hole*?

We're used to thinking of stars and black holes as just sitting out there. But really, they're objects in space, too, and they're also moving. In fact, everything in the Milky Way galaxy is moving around the center, and it's not like a nice and calm merry-go-round. It's entirely possible that in the scale of millions of years, an errant star or black hole could come our way.

That would be pretty disastrous.

If you set up a simulated solar system and shoot something at it with the mass of the Sun, it almost *always* ends in total catastrophe. The planets get tossed out into space. Sometimes, the black hole takes a planet with it as it leaves our neighborhood. What if a black hole comes and takes us away? Life in orbit around a black hole would be cold, dark, and short.

We don't see these things coming at us right now or in the next few thousand years, but over millions of years, it's totally possible.

It wouldn't be the first time our solar system got scrambled. If you look at our solar system over millions of years, you'll see that it's actually very chaotic. The solar system only *seems* like a calm and settled place to you now because we haven't seen it change in the last few hundred years. But on longer timescales, it can be a very dangerous place indeed. In fact, there's evidence all over the solar system of crazy catastrophes, such as the collision that led to the formation of Earth's moon or the weird gravitational event that gave Uranus its odd tilt. The solar system we see now is very different than the solar system we had billions of years ago.

If an errant rogue planet, star, or black hole came into our solar system, it's hard to imagine what future humans could do about it. Not even an army of Bruce Willises are probably capable of deflecting or destroying such a huge mass. At that point, we might only have one option left if we want to survive: take to the stars.

Billion-Year Threats

Let's look even deeper into the future. If humans survive for millions and millions of years, it's most likely because we've managed to populate the rest of the solar system or visit other stars. On those timescales, it's very likely that they've encountered something (an errant planet or black hole) that causes them to leave Earth.

But even if they haven't, we know that humans in the future will *have* to leave Earth eventually.

Our star, which has been burning happily for more than four billion years, is going to change. In about a billion years, it'll get much hotter, and grow *much* bigger. In fact, within a billion years, it'll be so big that its surface will be about where Earth is right now. So unless we develop some *really* amazing sunblock lotion technology, we're going to have to move. Maybe we'll move to the outer planets or to the asteroid belt. Remember Pluto? Let's hope it doesn't hold a grudge.

But even if we find a cozy asteroid or settle on Pluto, the clock will be ticking. After another billion years have passed, our Sun will fizzle out and retire, blowing out most of its gas and leaving behind a white dwarf that can't burn. What happens when the Sun cools and stops providing the warmth we need? Things are going to get . . . chilly. In order to survive beyond the next few billion years, it's clear that humanity will need to escape our solar system and travel to other stars.

Beyond

If there are still humans around billions or even trillions of years in the future, then it's a safe bet that they aren't on Earth, or even in this solar system. If we've somehow managed to survive that long, then it's pretty likely that we've learned to cross the vast distances of space and settled in other parts of the galaxy. In fact, if we've learned how to travel to other stars and colonize other planets, then there would probably be lots of human settlements all across the galaxy.

Imagine a human civilization spread across the galaxy. If we manage to get that far, would this mean that humanity has a chance to live forever?

After all, if you have humans across multiple solar systems, then you have a built-in insurance policy. Even if one solar system suddenly goes supernova, or if one human settlement goes astray and blows itself to smithereens, you still have other pockets of humanity to carry the torch (or the Nutella jar, as the case may be). Like a cockroach infestation, it would be pretty hard for the universe to wipe us *all* out at this point, wouldn't it?

And suppose we can do better than travel between the stars in our galaxy. What if humans in the future figure out how to cross the incredible distances between galaxies (via wormholes or fast-traveling ships), so that even if the Milky Way suddenly explodes or gets shredded to bits by crashing into another galaxy, some form of humanity still survives? Does that mean we have it made?

Not necessarily. At that point, there are still two big threats to human existence: the laws of physics and infinity.

Higgs Field Collapse

Some physicists believe that the foundations of the universe are not as solid as you might think.

For example, the mass of all matter particles could suddenly change, affecting how they move and interact with one another. This basic property is not fixed but comes from the particles' interaction with the energy stored in the Higgs field, which is one of the quantum fields that fills the universe. The problem is that physicists aren't sure how stable that field is. Someday, either spontaneously or triggered by some event, the Higgs field could collapse and lose its energy. If it does, the collapse would spread out across the entire universe, essentially scrambling all of physics. Such an event would probably destroy everything that we currently see in the universe, rearranging it into something totally different.

Scientists aren't sure how likely this is to happen or if it can happen at all. But on the scale of trillions of years and beyond, it's hard to predict what *can't* happen. And if it happens, there's basically no chance that any humans, even if they are spread across the stars, would survive.

Infinity

Infinity is a tough mother. Even if we manage to avoid all the things that can kill us, the sheer weight of time is going to get to us eventually. Infinity is a hard concept to grasp, but in an infinitely old universe, anything that *can* happen eventually *will* happen.

By spreading out across the stars, we might improve our chances of survival to 99.999999999999 percent, but over an infinite number of years, our time will come. Eventually, some fluke event that we can't foresee or imagine can (and will) happen, wiping out every human in existence.

So Are We Toast?

Before you feel too bad about the eventual demise of the human race, we should point out there is one way in which humans can live until the end of time. It's a bit of a technicality, but if we're imagining universe-faring humans eating Nutella in other galaxies, then now is not the time to hold back.

Imagine the scenario in which humans have figured out how to survive for billions and even trillions of years. And imagine that the weight of time or the Higgs field collapsing hasn't wiped us out yet. What if something unexpected happens? What if the universe stops expanding and instead suddenly reverses itself? What if that reversal causes the universe to come together and condense back into something really dense in a way that is the opposite of the Big Bang? Physicists call this the "Big Crunch," which coincidentally sounds like the name of a tasty, Nutella-filled candy bar.

If the Big Crunch were to happen, we would all be, well . . . crunched. Even if we see it coming, it's not something we can avoid or escape, because space itself will shrink. That means the universe gets smaller and smaller, so there's nowhere to run. If it goes far enough, space will contract to infinite density and then something super-weird is going to happen: time will end. It's going to end in the same way that, for example, heading north ends when you get to the North Pole. When you get there, there is nowhere more north. Similarly, when space and time crunch together, it's going to be the end of both.*

But imagine that we are still alive then, and imagine that we hang on until the very last gasp of the universe. Then technically, you *could* say that humans made it to the end of time. In fact, you could say they made it about as far as anyone could make it.

Wouldn't that be a victory in itself? To know that we lived to the fullest extent that we could live and that we used every second of time that we could have?

We should all be so lucky.

---

* At least, the end for this universe. Some physicists believe that the universe goes through cycles of Big Bangs and Big Crunches.

What Happens If I Get Sucked into a Black Hole?

A lot of people seem to have this question.

It's a common conundrum that gets covered in many science books, and it's a question that our listeners and readers often ask us. But why is that? Are there black holes popping up everywhere in backyards across America? Are there people out there who are planning to have a picnic near one and are worried about letting their kids run around it unsupervised?

Probably not. More likely, the fascination with falling into a black hole has less to do with the chances of it actually happening, and more to do with our basic curiosity about these intriguing space objects. And we get it: black holes are *mysterious*. They are weird regions of space out of which nothing can escape—voids in the fabric of space-time itself that are completely disconnected from the rest of reality.

But what would it be like to fall into one? Would you necessarily die? Would it feel different from falling into a regular hole? Would you discover deep secrets of the universe inside, or see time and space unfold before your very eyes? Would your eyes (or your brain) even work inside of a black hole?

There's only one way to find out, and that is to jump in. So grab your picnic blanket, say good-bye to your kids (maybe forever), and hang on, because we are about to take a dive into the ultimate backyard hazard.

Approaching the Black Hole

The first thing you might notice as you approach a black hole is that black holes do indeed look like . . . black holes. They are definitely black: black holes give off absolutely no light, and any light that hits them gets trapped inside. So when you look at one, your eyes don't see any photons, and your brain interprets that as black.*

They are also definitely holes. You can think of them as spheres of space where anything that goes into them stays there forever. What keeps things inside is the gravity of the things already in them: mass is compacted so densely in a black hole that the effects of gravity are enormous. Why? Because gravity gets stronger the closer you are to something with mass, and having mass compacted means you can get *really* close to it.

Typically, things with a lot of mass are fairly spread out. Take, for example, the Earth. The Earth has about as much mass as a black hole that's half an inch wide (about the size of a marble). If you were to stand a distance of one Earth radius away from the center of Earth and one Earth radius away from a marble-size black hole, you'd feel the same amount of gravity.

But as you get closer to each object, two very different things would happen. As you get close to the center of Earth, you will actually start to feel *no* gravity. That's

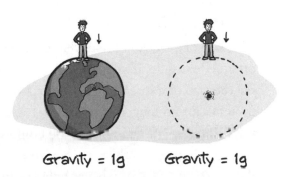

Gravity = 1g   Gravity = 1g

because Earth is now all around you, so you're being pulled equally in all directions. But as you get close to the small black hole, you'd

---

* Actually, black holes aren't *completely* black. They give off a tiny bit of radiation called "Hawking radiation" (named after Stephen Hawking), but it's so dim your eyes wouldn't register it.

feel an *enormous* amount of gravity. You'd feel the entire mass of Earth, all of it *really* close to you. That's what a black hole is: it's super-compacted mass, which makes it extremely powerful to the things immedi-ately around it.

A really compacted mass creates extreme gravity around itself, and at some distance, space gets so distorted (remember that grav-ity doesn't just pull on things; it distorts

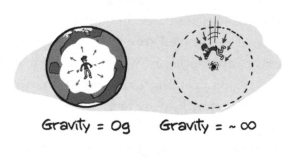

Gravity = 0g   Gravity = ~ ∞

space) that not even light can escape. The point at which light can no longer escape is called the "event horizon," and it (more or less)* defines where the black hole starts. It's the radius of the black sphere we call a black hole.

The size of a black hole can change, depending on how much mass you squeeze inside. If you compress Earth enough, you'll get a black hole the size of a marble because, at a distance of about a centime-ter, light can no longer escape. But if you add more mass, that dis-tance is larger. For example, if you compress the sun, the distortion of space is higher and the event horizon happens farther out, at three kilometers, giving you a black hole six kilometers wide. The more mass, the bigger the black hole.

In fact, there is theoretically no limit to how big a black hole can

---

* We say "more or less" because it's a little different for spinning black holes, and also because, as we'll later see, the black part is actually a little larger than the event horizon.

be. The smallest black hole we've detected in space is about twenty kilometers wide, and the largest is tens of billions of kilometers wide. Really, the only limitation is how much stuff is around to make the black hole and how much time you allow for the black hole to form.

The second thing you might notice as you approach a black hole is that black holes are often not alone. You would sometimes see stuff surrounding them, falling into the black hole. Or, more precisely, you would see the stuff swirling around waiting to fall into the black hole.

This stuff is called the "accretion disk." It's made of gas, dust, and other matter that didn't get sucked straight into the black hole but is instead circling in orbit, waiting for its turn to spiral in. With a small black hole it may not be very impressive, but with a supermassive black hole it can be a sight to behold. The sheer friction of all that gas and dust whipping around at ultrahigh speeds can be so intense that the matter gets ripped apart. This emits a lot of energy, creating some of the most powerful sources of light in the universe. These quasi-stars, or quasars, can sometimes be thousands of times brighter than all the stars in a single galaxy combined.

Fortunately, not all black holes, even supermassive ones, form quasars (or blazars for that matter, which are like quasars on steroids). Most of the time, the accretion disk doesn't have the right amount of stuff or the right conditions to create such a dramatic scene. This is a good thing because being close to a quasar would probably vaporize you instantly, way before you even get a glimpse of the black hole. Hopefully, the black hole that you picked to fall into has a nice, relatively calm accretion disk around it, and you actually have a chance at getting near it.

BLACK HOLE AND CHILL

Getting Closer

After you've confirmed that the black hole you're falling into doesn't have a swirling toilet bowl of burning gas and dust spraying more energy than billions of stars combined, the next thing you might want to worry about is death by gravity itself.

Usually when you hear the words "death by gravity," you think of falling to your death from something high, like a building or an airplane. But in those cases gravity isn't the one to blame. It's the landing that kills you, not the falling. In space near a black hole, however, it actually is the falling that can kill you.

You see, gravity doesn't just pull on you; it also tries to tear you apart. Remember that gravity depends on the distance to the object that has mass. When you stand here on Earth, your feet are closer to Earth than your head, which means your feet feel a stronger gravitational pull than your head. If you were to pull on one end of a rubber band harder than you are pulling the other end, the rubber band would stretch, even if you're pulling both ends in the same direction. That's what's happening to you right now: the parts of you closer to the ground feel more gravity and Earth is trying to stretch you like a rubber band.*

Of course, you probably don't feel stretched out, and that's because (a) our bodies are squishy, but not *that* squishy (i.e., we hold together pretty well); and (b) the difference in gravity between your head and your feet is not that strong. Gravity

---

* This is more applicable if you were to hop in the air, or be in free fall. When you're standing on the ground, your feet can't go anywhere, so gravity is actually trying to squish you flat.

on Earth is fairly weak, which means your head and your feet pretty much feel the same amount of gravity.

But if the gravity overall was much stronger, then you might be in trouble. If you were in free fall moving toward a really massive object, then the gravity might be strong enough for you to feel the difference in pull between your head and your feet. It's kind of like a playground slide: the taller the slide, the steeper it is on the way down. At some point, the difference in gravity between both ends of you might be enough to *actually* tear you apart.

This is where a lot of science books will tell you that surviving entering a black hole is impossible. They'll typically say that gravity around a black hole is so strong that you'd be "spaghettified" (aka pulled apart) before you even went in. But actually, this is not necessarily true! It's totally possible to enter a black hole.

It turns out that the point at which gravity would tear you apart (we'll call it the "spaghettification point") and the point at which light can't escape the black hole (i.e., the edge of the black hole) are not the same, and are actually in different places relative to each other depending on the mass of the black hole. The spaghettification point changes proportionally to the cubic root of the mass of the black hole, whereas the edge of the black hole changes linearly with the mass.

What this means is that for small black holes, the spaghettification point is bigger than the event horizon, which means it sits *outside* the edge of the black hole. But for large black holes, the spaghettification point is smaller and sits *inside* the black hole. For example, a black hole with the mass of a million suns has a radius of 3,000,000 kilometers, but its gravity won't pull you apart until you are deep inside, 24,000 kilometers from its center. On the other hand, a small black hole with a radius of 30 kilometers would

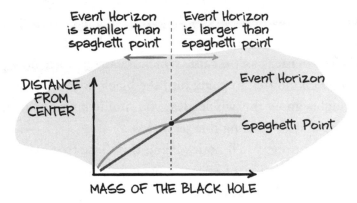

Event Horizon is smaller than spaghetti point

Event Horizon is larger than spaghetti point

DISTANCE FROM CENTER

Event Horizon

Spaghetti Point

MASS OF THE BLACK HOLE

pull you apart at a distance of 440 kilometers, way before you get to its edge.

It might be strange to think that smaller black holes are actually more dangerous to approach than bigger black holes, but that's just how the math of black holes works out. Bigger black holes cover such an enormous area that they don't need to be that powerful at their edges to suck things in and keep them inside.

Reaching the Black Hole

Okay, you've managed to pick a black hole that doesn't have a crazy party surrounding it, and one that is big enough that you won't be torn apart until *after* you get inside of it, which means . . . you're ready to go in. But watch out, this is when things start to get trippy.

As you get close to the black hole, you'll notice two interesting things.

First, at about three times the radius of the event horizon, you'll see that the accretion disk ends, leaving the area immediately surrounding the black hole largely empty. That's because any matter that gets closer than this point will quickly fall in. This is the point where most matter can't escape, which means you are now pretty much fully committed to going into the black hole. If you had second thoughts about this whole idea, you should have had them way before you started reading this section.

The second thing you'll notice being this close to a black hole is the enormous bending of space happening around you. You are now at a point where gravity is so strong that it distorts how light moves in very noticeable ways. It's like swimming inside of a lens: space around a black hole is so curved that light doesn't move in what seems like a straight line anymore.

Now we'll examine some of the wonky things that you will experience as you go further in.

The Shadow of the Black Hole*

At around two and a half times the radius of the black hole, you'll enter what is known as the "shadow" of the black hole. This is the actual black circle that anyone who looks at a black hole will see.

Black holes cast a bigger shadow than their size because they don't just capture photons that are inside of the event horizon; they also bend the photons flying nearby. Any light heading toward you that comes within a certain distance of the black hole will fall into the gravitational well and eventually go inside, which means you won't be able to see it.

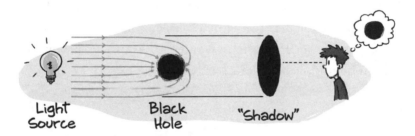

Light Source

Black Hole

"Shadow"

This shadow looks bigger as you move toward the black hole. As you get closer, the black hole captures even more light that would have otherwise hit your eyeball, which means the black hole will start to take up almost your entire field of vision.

Incidentally, this is the point where you want your friends to take your picture, because they'll see an image of you surrounded by pure blackness: it'll *look* like you are inside the black hole, even though you still have a ways to go.

---

* Also known as "The perfect title for that science fiction novel you've been wanting to write."

The Infinite Circle of Light*

At around one and a half times the radius of the black hole, you'll reach another fun milestone: the point at which light will orbit the black hole in a perfect circle. Just as planets and satellites can orbit around a more massive object, light can orbit around a black hole. What's remarkable about light orbiting a black hole is that light has no mass! This means that it's going around strictly due to the bending of space. A photon in orbit could potentially spin around a black hole forever, although any deviation will cause it to either spiral into the black hole or spiral outward into space.

One cool thing about crossing this point on your way to the black hole: because light goes in a perfect circle, when you look in any direction perpendicular to the black hole, you'll be able to see the *back of your head*. If you ever wanted to know what you looked like from behind, here is your chance.

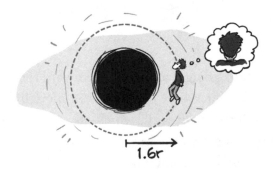

1.6r

---

* Also known as "The perfect title for that new age cult you've been meaning to start."

Bend It Like Beckham*

Closer than one and a half times the radius of the black hole, you've reached the point at which not even light can orbit safely. The chances of you escaping are dwindling, and all signs now point toward you going into the black hole, literally and figuratively.

By now, it will feel like the shadow of the black hole is enveloping you, closing off your view of the universe. If you look behind, you'll see an image of the universe shrinking.

What's weird about this view of the universe is that it will contain *all* of the universe, even the things that are behind the black hole. At this point, space is so bent that light comes from all sides of the universe, swirls around multiple times, and then hits the sides and back of your head. In this extreme, fish-eye view of the entire cosmos, you'll even see multiple copies of the universe repeated over and over at the edges of your field of vision.

As you move closer to the center of the black hole, this window out to the universe will shrink and shrink, and the image of the black hole will dominate everywhere you look.

And then . . . you'll cross the event horizon.

---

* Also known as . . . actually, someone already made this movie.

What Your Friends Will See

At this point, it's interesting to think about what your friends might make of all of this. You know, the friends who thought jumping into a black hole was crazy and stayed behind? We're sure they were very supportive of you, but what do *they* see as you take this glorious leap into the unknown?

It turns out they never see it happen. Not because it gets obscured by the darkness of the black hole, but because it literally never happens *for them*.

Remember that gravity doesn't just distort space; it also distorts *time*. And black holes have so much gravity that they distort time in a really extreme way.

A lot of people know that time slows down at very high speeds. For example, if you climb onto a spaceship and zoom off at nearly the speed of light and come back, time will have moved slower for you and everyone you know will have aged more than you. But it's not just speed that can have this effect on time; being near a really massive object (like a black hole) does this, too. It bends space, but it also slows down time.

As you dive near the black hole, your friends will see time slow down for you. To them, you'll start to look like you are traveling in suuuuuper . . . slooooooow . . . moooootion. They'll see you get closer and closer to the black hole, but you'll be doing it slower and slower.

And the closer you get to the black hole, the slower your clock will go. At some point, your clock will slow down so much that, to them, you will almost look like you are frozen in time. We're sure they're great friends, but eventually they'll probably give up and go live the rest of their lives. The last image they'll have of you will be faint and red, because the gravity also stretches the wavelength of the photons into the infrared spectrum.

In fact, it won't just be a long time to the rest of the universe before you fall in. It will literally never happen. From an outside perspective, time for you will freeze, and your image will get spread across the surface of the black hole and be forever etched there. It would take an infinite amount of time before an observer sees you fully enter the black hole. Solar systems and galaxies would form and perish. Trillions of years would pass, and they would never see you cross the boundary.

If you were hoping to impress your friends with a dramatic move, jumping into a black hole was not a wise choice.

Entering the Black Hole

Of course, that's just what your friends see. To you, it's still a wild roller coaster ride.

Remember that time still moves normally for *you*, so from your perspective the trip into the black hole will happen at normal speed.

You *will* go into the black hole; it's just that, to the universe outside, it will appear as if it never happens.

So what happens when you finally cross the event horizon? Not much, physicists believe.

As you pass over the final threshold, your view of the outside universe shrinks to a smaller and smaller dot and everything else around you becomes completely dark. The only source of light you can see is that dot exactly behind you that contains a tiny picture of the entire universe. So that's something. But according to theory, there's nothing actually *at* the event horizon. There isn't a wall or a fence or a force field or confetti, or a gate staffed by galactic security guards. It's just the place in space where you have no way to return.

You see, inside the black hole space is bent so much that there are no paths out. No matter how fast you go, space-time has become one-directional. Outside the black hole, only time was one-directional (forward). But inside the event horizon, space is also one-directional (inward). Every trajectory inside the black hole leads deeper in.

This change would have been gradual, not sudden, for you. As you were coming closer to the event horizon, the possible paths you

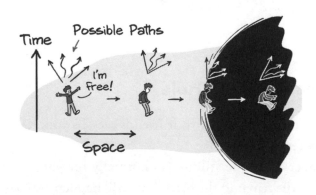

could take started getting distorted, too. There were fewer and fewer paths away from the black hole. The event horizon is just the point at which all possible paths available to you point inward.

One thing is clear: you are definitely stuck now. Escape at this point is worse than futile. If you struggle and try to flee, you will only move faster toward the center of the black hole.

What's Inside?

So what's it like, now that you are inside the black hole?

The truth is that nobody knows. In fact, we may *never* know.

We don't even know if it's even possible to *think* inside a black hole. Our bodies require the movement of blood and information and ions in all directions. If your neurons and your blood can only fire and flow toward the center of the black hole, would you even be alive, much less conscious?

More fundamentally, though, we don't really know what space and time are like beyond the event horizon. We have an *idea* about what happens. The theory of general relativity has so far been right about everything that happens outside of black holes (even predicting their existence), but we also know that general relativity isn't the truest description of how the universe works. We know, for example, that it breaks down at the smallest level, where quantum mechanics can't be ignored. So is it possible that general relativity breaks down inside a black hole? Most definitely, but we aren't sure how much it might be wrong, or whether it might only be wrong at the very center of the black hole.

If general relativity is still mostly right inside of the black hole, then what happens next is not that exciting. According to general relativity, gravity will just continue to get more intense, and you will move toward the center of the black hole faster and faster. In fact, for a black hole like the one at the center of the Milky Way, you would fall to the center in about twenty seconds. Of course, you would never make it to the center, because you will most definitely hit the spaghettification point (remember that?) at some point and get shredded to bits.

But if general relativity *isn't* right about what happens just inside the event horizon, then we are free to speculate about what could be happening. It turns out, a whole host of fun things could be waiting for you when you go in:

✦ **Another universe.** Some physicists think (and even say it's likely) that a whole other universe could exist inside of a black hole. Perhaps when you go into the black hole, you'll pop out at the beginning of a new baby universe.

✦ **A wormhole.** Another theory is that the inside of a black hole can connect to a wormhole (a sort of tunnel in space-time), taking you to another part (and time) of the universe. What's at the other end? Scientists speculate that on the other side you could get spit out by the opposite of a black hole: a *white* hole. If a black hole is a place where things can enter but never escape, then a white hole is a theoretical place where things can escape but never enter. Think of a white hole as a region of space where the space is bent in such a way that all directions point you out of the white hole. Of course, you might be thinking,

Where do the things that come out of the white hole come from? They come from the black hole through the wormhole!

THINGS YOU MIGHT FIND INSIDE A BLACK HOLE:

CERTAIN DEATH   ANOTHER UNIVERSE   A WORMHOLE   EINSTEIN CHILLIN' WITH SCHRÖDINGER

In either case, this would be the end of your journey, at least from the perspective of our universe. Once inside the black hole, it's very unlikely that you'll ever get out, which means that whether you die a horrible death, discover the secrets of quantum mechanics and general relativity, or find a whole new universe, only *you* will know this fabulous secret.

The only problem is, you won't be able to tell anyone.

Why Can't We Teleport?

L et's face it: nobody likes to travel.

Whether they're traveling to get to an exotic location for vacation or traveling to work on a daily commute, nobody actually likes the part where they have to travel. The people who say they like to travel probably mean they like to *arrive*. That's because being somewhere can be really fun: seeing new things, meeting new people, getting to work sooner so you can go home early and read physics books. The actual *traveling* part is usually a drag: getting ready, rushing, waiting, rushing some more. Whoever said "it's the journey, not the destination" clearly never had to sit in traffic every day and never got stuck in a middle seat on a transatlantic flight.

Wouldn't it be great if there was a better way to get to places? What if you could just *appear* where you want to go, without going through all the places in between?

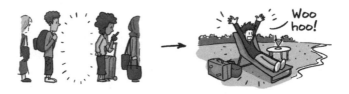

Teleportation has been a fixture in science fiction for well over a hundred years. And who hasn't fantasized about closing their eyes or hopping into a machine and suddenly finding themselves where they want to be? Think of the time you'd save! Your vacation could start *now*, and not after a fourteen-hour flight. We could get to other planets more easily, too. Imagine sending colonists to the nearest habitable planet (Proxima Centauri b, four light-years away) without having to spend decades in transit.

But is teleportation possible? And if it is, why is it taking scientists so long to make it a reality? Will it take hundreds of years to develop, or can I expect it as an app on my phone sometime soon? Set your phasers on stun, because we are going to beam you up on the physics of teleportation.

Options for Teleportation

If your dream of teleportation is to be here in one moment and then be in a totally different place the next moment, then we are sad to tell you right off the bat that this is impossible. Unfortunately, physics has some pretty hard rules about *anything* happening instantaneously. Anything that happens (an effect) has to have a cause, which in turn requires the transmission of information. Think about it: in order for two things to be causally related to each other (like you disappearing here and you appearing somewhere else), they have to somehow talk to each other. And in this universe, everything, including information, has a speed limit.

Information has to travel through space just like everything else, and the fastest *anything* can travel in this universe is the speed of light. Really, the speed of light should have been called the "speed of information" or "the universe's speed limit." It's baked into relativity and the very idea of cause and effect, which are at the heart of physics.

Even gravity can't move faster than light. Earth doesn't feel gravity from where the Sun is *right now*; it feels gravity from where the Sun *was* eight minutes ago. That's how long it takes information to travel the ninety-three-million miles between here and there. If the Sun disappeared (teleporting off for its own vacation), Earth would continue in its normal orbit for eight minutes before realizing that the Sun was gone.

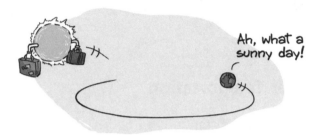

So the idea that you can disappear in one place and reappear in another place instantly is pretty much out of the question. Something has to happen in between, and that something can't move faster than light.

Fortunately, most of us aren't such sticklers when it comes to the definition of "teleportation." Most of us will take "almost instantly" or "in the blink of an eye" or even "as fast as the laws of physics will allow" for our teleportation needs. If that's the case, then there are two options for making a teleportation machine work:

1. Your teleportation machine could *transmit* you to your destination at the speed of light.

2. Your teleportation machine could somehow shorten the distance between where you are and where you want to go.

Option #2 is what you might call the "portal" type of teleportation. In movies, it would be the kind of teleportation that opens up a doorway, usually through a wormhole or some kind of extradimensional subspace, that you step through to find yourself somewhere else. Wormholes are theoretical tunnels that connect points in space that are far away, and physicists have definitely proposed the existence of multiple dimensions beyond the three we are familiar with.

Sadly, both of these concepts are still very much theoretical. We haven't actually seen a wormhole, nor do we have any idea how to open one or control where it leads. And extra dimensions aren't really something you can move into. They only represent extra ways in which your particles might be able to wiggle.

Much more interesting to talk about is Option #1, which, as it turns out, might actually be something we can do in the near future.

Getting There at Light Speed

If we can't appear in other places instantly, or take shortcuts through space, can we at least get there as fast as possible? The top speed of the universe, three hundred million meters per second, is plenty fast to cut your commute down to a fraction of a second and make trips to the stars take years instead of decades or millennia. Speed-of-light teleportation would still be awesome.

To do that, you might imagine a machine that somehow takes your body and then pushes it at the speed of light to your destination. Unfortunately, there's a big problem with this idea, and it's that you're too heavy. The truth is that you're too massive to ever travel at the speed of light. First, it would take an enormous amount of time and energy just to accelerate all the particles in your body (whether assembled or broken up somehow) to speeds that are close to the speed of light. And second, you would never get to the speed of light. It doesn't matter how much you've been dieting or working on your CrossFit; nothing that has any mass can ever travel at the speed of light.

Particles like electrons and quarks, the building blocks of your atoms, have mass. That means that it takes energy to get them moving, a lot of energy to get them moving fast, and *infinite* energy to reach the speed of light. They can travel at very high speeds, but they can never achieve light speed.

That means that you, and the molecules and particles that make up who you are right now, would never actually be able to teleport. Not instantaneously, and not at the speed of light. Transporting your body somewhere that quickly is never going to happen. It's just not possible to move all the particles in your body fast enough.

But does that mean teleportation is impossible? Not quite!

There is one way it can still happen, and that's if we relax what "you" means. What if we didn't transport you, your molecules, or your particles? What if we just transmitted the *idea* of you?

You Are Information

One possible way to achieve speed-of-light teleportation is to *scan* you and send you as a beam of photons. Photons don't have any mass, which means they can go as fast as the universe will allow. In fact, photons can *only* travel at the speed of light (there's no such thing as a slow-moving photon).*

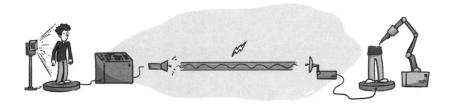

Here's a basic recipe for speed-of-light teleportation:

Step #1: Scan your body and record where all your molecules and particles are.

Step #2: Transmit this information to your destination via a beam of photons.

---

* In a vacuum.

Step #3: Receive this information and rebuild your body using new particles.

Is this possible? Humans have made incredible progress in both scanning and 3D printing technologies. These days, magnetic resonance imaging (MRI) can scan your body down to a resolution of 0.1 millimeters, which is about the size of a brain cell. And scientists have used 3D printers to print increasingly more complicated clusters of living cells (known as "organoids") for testing cancer drugs. We've even made machines (using scanning tunneling microscopes) that can grab and move individual atoms. So it's not hard to imagine that one day we might be able to scan and then print whole bodies.

The real limitation, though, might not be technological but *philosophical*. After all, if someone made a copy of you, would it actually be *you*?

Remember, there's nothing particularly special about the particles that make up your body right now. All particles of a given type are the same. Every electron is perfectly identical to every other electron, and the same is true for quarks. Particles don't come out of the universe factory with personalities or any sort of distinguishing features. The only difference between any two electrons or any two quarks is where each of them is and what other particles they're hanging out with.*

But how much would a copy of you still be you? Well, it depends on two things. The first is the resolution of the technology that scans and prints you. Can it read and print your cells? Your molecules? Your atoms, or even your individual particles?

The even bigger question is how much your "you-ness" depends on the tiny details. What level of detail does it take for the copy to still be considered *you*? It turns out that this is an open question, and the answer might depend on how quantum your sense of self is.

A Quantum Copy of You

How much information would have to be recorded in order to create a faithful copy of you? Is knowing the location and type of every cell and connection in your body enough? Or do you also need to know the position and orientation of every molecule in your body? Or if you drill down deeper, do you also need to record the quantum state of every particle?

Every particle in your body has a quantum state. That quantum state tells you where the particle is likely to be, what it's likely to be doing, and how connected it is to other particles. Because you can only say what each particle is *likely* to be doing, there's always some uncertainty. But is that quantum uncertainty an important

---

* Electrons are actually just little self-sustaining bundles of energy in the quantum fields that fill space. When an electron moves, it means the field in its old location stops buzzing and the field in its new location starts to buzz. So at the quantum level, every motion of a particle could be considered teleportation!

part of what makes you *you*? Or does it happen at such a small level that it doesn't really influence important things, like your memories or how you react to things?

At first glance, it seems unlikely that the quantum information in each of your particles would make a difference in making you who you are. For example, your memories and your reflexes are stored in your neurons and their connections, which are pretty big compared to particles. At that scale, quantum fluctuations and uncertainty tend to average out. If you were to subtly scramble the quantum values of a few of the particles in your body, would you be able to tell the difference?

Debating the answer to this question might be more appropriate for a philosophy book, not a physics book, but here we can at least consider the possibilities.

You're Not That Quantum

If it turns out that the quantum state of your particles doesn't play a role in making you who you are, and that simply re-creating how your cells or molecules are arranged is enough to make a copy that thinks and acts like you, then this is good news for your next vacation, because teleportation gets *a lot* easier. This means that you just have to record the location of all your small bits and pieces, and then put them together in the exact same way elsewhere. This is like taking a LEGO house apart, writing out the instructions, and then sending those instructions to another person to build. Modern technology seems to be well on its way to someday achieving this.

Of course, it wouldn't be an *exact* copy of you, which might make you wonder if you're losing something in the translation.

Would it be like sending a JPEG version of an image instead of the full picture? Would you come out the other end a bit fuzzy around the edges, or not feeling quite like yourself? The loss of fidelity you're willing to put up with depends on how badly you want to get to the next star system in as short a time as possible.

You're Totally Quantum

But what if your you-ness *does* depend on quantum information? What if the magic, or the indelibleness of you, lies in the quantum uncertainty of every particle in your body? This sounds like a bit of new age hocus-pocus, but if you really want to be certain that the copy coming out the other end of this teleportation machine is *exactly* the same as you, then you have to go quantum all the way.

The bad news is that this makes the problem of teleportation *much* harder. Really, anything quantum is hard, but the idea of copying quantum information is doubly hard.

This is because, from a physics point of view, it's technically impossible to know everything about a particle all at once. The uncertainly principle tells us that when you measure the position of a particle very accurately, you can't know the velocity, and when you measure the velocity, you can't know the position. And it's not just that you can't know it. It's much deeper: information about position and velocity *doesn't simultaneously exist*! There's an inherent uncertainty in every particle.

The only thing you can know about a particle is the *probability* that it's here or there. How, then, do you make a quantum copy with the same probabilities as the original?

Making a Quantum Copy

Let's consider the problem of making a quantum copy of a single particle. If you insist that your light-speed teleportation machine make a copy of you that is absolutely identical to your current self, then this is pretty much your only option.

To copy a particle down to the quantum level means that you want to copy its quantum state. The quantum state of a particle includes the uncertainty about its position and velocity, or about its quantum spin, or any other quantum property. It's not really a number but more a set of probabilities.

The problem is that to extract quantum information from a single particle, you have to probe that particle somehow, which means perturbing it. Even just *looking* at something involves bouncing photons off of it. If you shoot photons at an electron, you might learn about its quantum state, but you will also scramble it. This isn't because we aren't clever enough or because we haven't developed a fine enough probe. The quantum "no-cloning" theorem tells us that it's *impossible* to read quantum information without destroying the original.

So how do you copy something that you can't see or touch? It's not easy, but one way to do it is using "quantum entanglement." Quantum entanglement is a strange quantum effect where the probabilities of two particles get linked together. For example, if two particles interact with each other so that you don't know what their spins are, but you do know that they are the opposite of each other, then the two particles are said to be entangled. If you find that one is spinning up, you know the other one must be spinning down, and vice versa.

Quantum teleportation works by taking two particles, entangling them, and then using them like two ends of a telephone fax line. For example, you can take two electrons, entangle them, and then send one of them to Proxima Centauri. Those two electrons would sit there, still entangled, until you are ready to start the copy process.

From there, it gets a little complicated, but essentially you use the entangled electron you have here to probe the particle you want to copy, and that interaction gives you the information you need to make the electron at Proxima Centauri be an exact quantum copy of the particle you wanted to duplicate.

Step
①

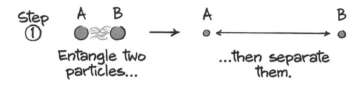

Entangle two
particles...

...then separate
them.

Step
②

Entangle one of the particles you separated
with the particle you want to copy.

Step
③

Peek at the quantum state
without destroying it.

Step
④

Share what you saw with
someone on the other side.

That information lets them convert the
second particle to a quantum copy.

Amazingly, humans have done this for single particles and even for small groups of particles.* The record so far is making a quantum copy between two points that are 1,400 kilometers apart. That won't get you to Proxima Centauri yet, but it's a start.

Scaling this quantum copy machine to more than just a few particles won't be easy. There are 10^{26} particles in your body, so it gets very complicated, very fast. But the point is that it's *possible*.

Is that quantum reassembled person *actually* you? Well, it would be the most faithful reproduction of you that can possibly be made. If that's not you, then who are you?

Too Many Yous

One potentially sticky part about this idea of teleportation is that it can end up making multiple copies of you. In the case of the low-fidelity teleportation machine that doesn't copy quantum information, you might imagine using it to make clones of you. You

---

* Note that, as cool as that is, quantum teleportation won't let you do anything faster than light, since it still requires you to share what you saw using normal communication, which is limited to light speed.

could scan your body and then beam that information to Proxima Centauri, and then to Ross 128 b (another nearby habitable planet), and then to any number of other planets. In fact, you could start printing copies right here. They might not be exact quantum copies of the original, but they would be similar enough to create all kinds of moral and ethical issues.

Fortunately, there is one saving grace about the quantum copying version of the teleportation machine. The same principles of quantum theory that allow you to copy quantum information also require that the original information be destroyed when it's copied. Whichever way the technology ends up working, the scanning process would inevitably destroy the original by scrambling all of its quantum information. This means that the copy you send over is the only copy that remains.

Beam There, Done That

To recap, the idea of transporting ourselves somewhere in a proverbial blink of an eye is definitely possible. If you can tolerate a speed-of-light transmission delay, and if you accept that a scanned and reassembled version of you is really you, then teleportation just might be in your future.

It's for my NEXT vacation.

Of course, we forgot one important caveat: in order to teleport somewhere as described in this chapter, there needs to be a machine on the other side to receive your signal and reconstruct you. That means that if you want to one day beam yourself to another planet, someone has to first get there the old-fashioned way: by traveling.

Any volunteers?

Is There Another Earth Out There?

t's always good to have a backup.

Did you spill your coffee on your pants at work? Just pull out that spare pair you keep in your pants drawer at your desk. Did your kid lose their favorite special stuffed animal just before bedtime? You'll be glad you bought five identical copies on that IKEA trip.

Life in this crazy random universe can be pretty unpredictable, so it makes sense to keep a spare copy of the things that are important to you. And the more important it is, the more effort you should put into having a backup, right? So it shouldn't be a surprise that several of our listeners have written to ask if there's another, backup planet Earth out there. You know, just in case.

Of course, spilled coffee is not the kind of thing that would require us to move our entire civilization off to another planet. But it's still a valid point. After all, there are lots of real reasons why we might need a new home.

For example, what if we find that a giant, planet-killing asteroid is headed toward Earth? Or what if our robot vacuum cleaners one day get tired of cleaning up after us and decide to take over and kick us out? Or what if a supernova explodes nearby, blasting Earth with deadly radiation and killing every human on it? Clearly, it would be a good idea to have another planet we could call home. Otherwise, we are literally putting all our eggs in one basket.

But how easy would it be to find a second home? Did we luck out with Earth or are there plenty of cozy, livable planets out there in the cosmos? Let's pretend we have all the money in the world and go on the ultimate house-hunting mission.

Cosmic Neighbors

Everyone who keeps an extra pair of pants in their desk (and who doesn't?) knows they do it for a reason. When you need a spare, you want it to be close by. In the same way, it would be great if we could find another planet to live on right here in our solar system. If anything were to happen to Earth, it would save us a lot of hassle if we could just hop over to our new house without having to pack for a centuries-long space trip.

Unfortunately, there aren't that many great options here in our solar system.

Let's start with our closest neighbor, Venus. Venus is pretty much a nonstarter. The temperature on Venus is more than 800°F on the surface, and it has an atmospheric pressure ninety times higher than Earth's. In other words, Venus is not a good backup in case of disaster.

Our other closest neighbor is Mars. Mars is beautiful, and even looks a little like the Arizona desert on a hazy day. But Mars is also

not a great option for us to live on. Scientists think that Mars once had a planetwide magnetic field, just like Earth, but at some point lost it. We're not sure why, but perhaps it was due to its molten core cooling. Few people realize how important having a magnetic field is: it basically acts as a force field to protect us from the Sun's deadly solar winds. Without it, you not only get bombarded by deadly radiation but your atmosphere gets blown away, which is a big problem. If you don't have an atmosphere, you can't hold on to any heat, which means it gets *really* cold. Mars is another worst-case scenario for what can happen to our planet.

Beyond these two planets, things don't get any better. Mercury, which is just past Venus, is pretty bad. It's only fifty-seven million kilometers from the Sun, and it barely rotates, which means one side of it is always fried to a crisp and the other side is always frozen solid. It's the planetary equivalent of Baked Alaska: great for a dessert, not so good for housing billions of cosmic refugees.

Looking away from the Sun, our options don't improve much.

The planets beyond Mars are either too dark, too gassy, or just too frozen.

Jupiter and Saturn are basically enormous balls of gas. Even if you could survive their atmospheres of mostly hydrogen and helium, there wouldn't be anywhere to stand. Their solid cores are deep within the planets, under enormous pressures, and they are made up mostly of metallic hydrogen.

Neptune and Uranus, the farthest planets from the sun, are also no picnic. These planets are called "ice giants" because, well, they are giant balls of ice. Moving to one of these two planets would make as much sense as building a summer house in Antarctica.

Some scientists who look at the orbits of small objects past Neptune and Uranus think they see weird patterns that suggest there might be another planet hidden out there. They call it "Planet X." Unfortunately, even if it did exist as a planet (other scientists think it could be a dark matter blob, or even a black hole left over from the Big Bang), it would also be too cold.

What about any moons in our solar system? Are there any decent-size *moons* out there where we could live? Jupiter and Saturn are so large that some of their moons are as big as some of the inner planets. Sadly, most of these are also frozen solid. There is one moon of Jupiter, Io, which has hot volcanoes. But on Io you have to make a choice between the frozen surface (–202°F) or the fiery volcanoes (3000°F). There's no happy middle.

So as we look for a cosmic second home, it seems like there aren't any good options here in our solar system. We seem to be in the undesirable real-estate situation of having the best house on the block. That means it's time to search beyond the planets in our neighborhood and look at the rest of the universe.

Planets Beyond Our Solar System

For a long time, we didn't know whether there were many planets outside our solar system, or if our Sun was the only star to have them. All of the great thinkers in history, from Plato to Newton to Galileo to Einstein and Feynman, looked up at the sky wanting to know the answer to this question. Unfortunately, they all died not knowing the answer, which we only learned for certain about twenty years ago.

Think for a moment about how lucky you are. You are alive at this moment, just as we are discovering what is truly out there in the universe. Today, humans have figured out how to detect and even *see* planets around other stars, and it turns out that the answer to that ages-old question is: there are *lots* of planets out there. So many planets.

For thousands of years, humans thought there was only one planet in existence: Earth. It was a long time before we even thought that there could be other planets. One of the earliest mentions of this idea came from the ancient Babylonians, who knew of six planets, out to Jupiter, and left clay tablets recording their motion more than three thousand years ago. Progress was pretty slow for a long time, until the invention of the telescope.

The telescope let early scientists study the stars and think more clearly about how similar they were to our Sun. And if our Sun had so many planets, maybe other stars could have them, too. When we started to grasp the sheer size of our galaxy, and the huge number of stars that were in it, the number of possible other planets exploded. Astronomers got the sense that the number of planets in our galaxy could be numbered in the hundreds of billions.

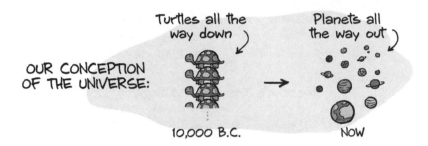

OUR CONCEPTION OF THE UNIVERSE:

Turtles all the way down

Planets all the way out

10,000 B.C.

NOW

Then in 1995, scientists finally started to see these planets. By looking at how the light from stars changes in frequency, they figured out how to tell if a particular star was being pulled around by the planets orbiting around it. This was a monumental achievement. It meant we could check for planets without actually needing to see them directly, which is hard to do.

In 2002, we figured out another clever way to detect planets. If a star has a planet orbiting around it, and if that planet passes between us and that star, we can actually see the light from the star dip as the planet blocks it from view. This is what the Kepler telescope has been doing for the past few years: it takes pictures of thousands of stars and it looks for dips that tell us which ones have planets.

We've also made progress in being able to directly see planets around other stars. It's almost an impossible task, because stars are so far away and they are extremely bright compared to any planets orbiting around them. Seeing a planet around a distant star is like standing in Los Angeles and trying to see a small candle that's next

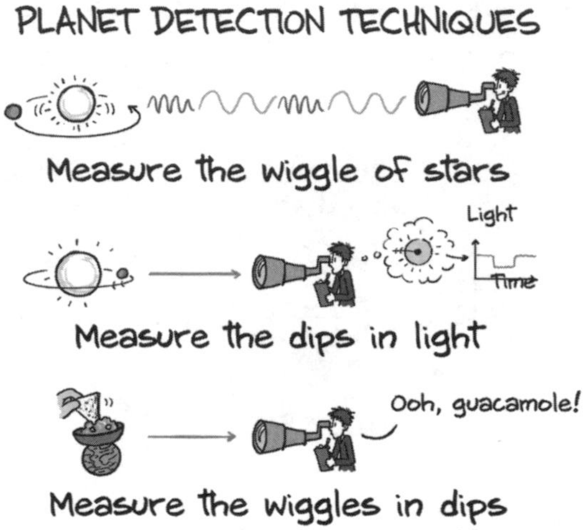

PLANET DETECTION TECHNIQUES

Measure the wiggle of stars

Light

Time

Measure the dips in light

Ooh, guacamole!

Measure the wiggles in dips

to a giant lighthouse in New York. And yet astronomers have done it: humans have actual, though fuzzy, pictures of other planets.

All of these techniques have exploded our ability to detect other planets. We went from knowing there are ~~nine~~ eight planets in our solar system to actually having data about thousands of planets.

What we learned is that the universe is *teeming* with planets. Just in our galaxy, we think that there are hundreds of billions of them. Picture every star in the night sky, and then imagine several planets orbiting around each and every one of them.

All of this might make you think that we have lots of options for finding a second Earth. But how many of those planets are actually good for us to live in? What are the chances that any of them will be as cozy and comfortable as the one we currently have?

A Good Home

If you are going to go through the trouble of packing everything up and settling on another planet, there are a few things you might want to check before calling the movers. After all, you don't want to pick a planet and then discover when you get there that it doesn't have enough bathrooms for everyone. Following is a list of things to look for when going planet shopping.

Proximity

We think that on average stars have around ten planets, which means there must be trillions and trillions of them in the universe. And if the universe is infinite, then there might even be an *infinite* number of planets out there. But realistically, how many of them could we get to? The nearest galaxy to us (Andromeda) is about 2.5 million light-years away. If the thought of sitting in your car with your kids for 2.5 million years sounds less than appealing, you may want to limit your options to just the planets in the Milky Way, which is a more manageable 100,000 light-years wide.

Rockiness

If you go to a lot of planet open houses, you'll soon discover that they come in basically two models: with and without rocks. Rocky planets are, obviously, mostly rock, and they have all sorts of advantages, like being able to stand on them and walk around. The other kind of planet is a gas planet, which offers fascinating things like hundred-year-old, Earth-size, violent storms, but lacks basic amenities such as having a place to land your spaceship, or . . . any land at all.

How many rocky planets are there? Fortunately, there are a lot! Scientists have learned that most of the stars in the galaxy have on average at least one rocky planet. This is good news for those of you who like having your house on solid ground because it means there are at least one hundred billion rocky planets in the Milky Way Galaxy. They range in size from Earth size to super-Earth size (up to fifteen times the size of Earth).

The Goldilocks Zone

Before you start celebrating all of our options for a second home, think a bit more carefully about what it would be like to live on a random rocky world out there. Some of those planets might be really close to their stars, which means you'd get blasted by solar radiation and fried to a crisp like on Mercury. Or they might be orbiting so far away that if you stood on them and looked up, the sun would just look like any other star, shining over a frozen ball of lifeless rock.

If you're going to pick a planet to live on, you want it to be not too close to its sun and not too far, so that your planet doesn't get too hot or too cold. Scientists have a perfect name for this prime area of real estate: the "Goldilocks zone." Interestingly, the Goldilocks zone is not the same for every star. For a superhot giant star, a comfortable distance would be really far from it. And for a cold, dim star, you want to get much closer to avoid freezing. Most of the stars in the galaxy (around 70 percent) are of the smaller variety (called M dwarf stars) and are usually much dimmer than ours.

Surprisingly, picking only planets that are in the Goldilocks zone of their stars only reduces the number of possible planets we can colonize by about a factor of two, because most rocky planets are close to their suns anyway.

Oh Yeah, an Atmosphere

This is starting to sound easy, right? You can just imagine lying by a pool on your new planet, feeling good, and taking a deep breath of . . . what? Oops, we forgot to check for an atmosphere.

We're used to being able to breathe the air here on Earth, but we often forget how lucky we are. Not every planet has a superthin layer of gas that helps make our kind of life possible. That's because atmospheres are rare and they are all too easy to lose. Most of Earth's atmosphere was made by volcanic eruptions in the early days of our planet. Think of the air you breathe as the result of geological indigestion. But not every planet goes through this process. And even if they do, they often lose all their burps into space. Radiation from space (usually the sun) is constantly trying to blow it off the planet like wind on a cheap toupee.

But how can we tell which rocky planets in a Goldilocks zone also have atmospheres? It'd be a bummer to travel that far only to choke once we get there. Fortunately, scientists have also figured out how to check the atmosphere of faraway planets. You might think it's impossible, given that we barely have fuzzy, pixelated views of them, but once again the secret is in the light.

When a planet is in front of its star, it blocks some of the star's light. But a tiny sliver of the light passes through the atmosphere of the planet, which changes the color of the light. Just like a sunset or sunrise on Earth makes the light from the Sun look more red, sunsets and sunrises on planets around other stars give us clues about whether their atmospheres would be wonderful and fresh or instantly melt our lungs with acid.

Amazingly, we can even tell the weather in some distant planets. By looking at how the atmosphere changes as the planet goes around its star, we can infer things like air currents and temperature. And it works! Atmospheres have been spotted around distant planets, and very recently astronomers found a mini-Neptune (a *Neptunito*?) about 120 light-years away that has the characteristic light signature of water vapor. Water in the atmosphere means there might also be water on the surface, and even oceans. So pack your swim trunks!

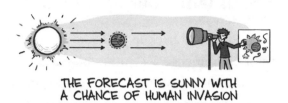

THE FORECAST IS SUNNY WITH
A CHANCE OF HUMAN INVASION

Of course, you don't just want the warm blanket of an atmosphere; you also want it to not kill you instantly when you breathe it in. It would be great if the atmosphere in our new home had all the fresh air components of our atmosphere here on Earth. Unfortunately, oxygen in its breathable O_2 form seems to be rare in the universe. On Earth, we only have it because a huge number of microbes developed photosynthesis, producing oxygen gas as a byproduct. This process took billions of years on Earth, which is a lot longer than we'd like to wait to move into our new home. So if we are to find a new home, we need to find a planet where that process already started a billion years ago, which means we need to find a planet with life (microbial life, that is) already in it. It's almost the opposite of how we look for houses here. On Earth, nobody wants to buy a house that's full of bacteria, but for a second planet, you hope to find one where they've taken over!

Pack Your Bags (and Your Spare Pants)

In summary, to find a good backup planet to live on, we need to be way pickier than Goldilocks. We know there are billions and billions of rocky planets in cozy spots in our galaxy, but how many of them have protective atmospheres and bacteria that can make breathable oxygen? The science of finding atmospheres and life on alien worlds is still too new to give us good estimates for the number of planets that have them. But the fact that we've already found a few planets with atmospheres (and even some with possible signs of life) tells us that perhaps it's not impossible.

While there might be cozy, Earthlike planets out there, there is still the question of whether we could get to them. Even if we find another perfect Earth on the other side of the galaxy, we still have to make the trek there, which is a daunting 100,000-light-year schlep. We have no idea if we can make a trip that far, or even survive in space that long. The Earth we have right now might be the only Earth we'll ever get.

So until warp drives or wormholes become a reality, keep an eye on your robot vacuum and, for goodness sake, try not to spill your coffee.

What's Stopping Us from Traveling to the Stars?

T raveling to the stars would be *really* exciting. We get excited just writing the phrase "to the stars!" To finally break out of our little planetary prison and explore the cosmos would be a huge milestone for humanity.

HUMANITY BUCKET LIST

☑ Invent Smartphone   ☑ Make 11 Star Wars movies   ☐ Travel to the stars

For our entire history, we've been confined to one small corner of space. With the exception of the twelve astronauts who stepped on the moon, all of the billions of humans that have ever walked the earth have been trapped on this tiny, rocky home.* And even those twelve that escaped Earth's gravity barely got to explore our cosmic neighborhood. Their short hop to the moon was the galactic equivalent of leaving the house and visiting the garage.

And yet we know there is a lot more out there to explore and to experience.

Our telescopes have given us a wide and far-reaching view of the universe. We can see distant stars and galaxies, and we know there are countless numbers of them. We even have images of other *planets* orbiting those stars and hints about what it would be like to live there. The explorer inside all of us is driven crazy with curiosity: What are those planets really like? Could they be potential future homes for humanity? Are there aliens living in them who could share with us deep secrets of the universe? Traveling to the stars would let us answer all of these questions and more.

But the fact remains, we've hardly even left our solar system.† What exactly is keeping us from exploring the cosmos? Are there actual physical laws that prevent it, or is it just a matter of devel-

---

* As of this writing only four of those twelve moon-walkers are still alive. So the chances that you, dear reader, have stepped off Earth are approximately four in one hundred billion.
† The Voyager 1 probe left the solar system (or, more accurately, the heliosphere) in 2012.

oping the right technology? Let's take a look at the challenges that make space travel a difficult proposition.

It's a *Big* Universe

As we learned in previous chapters, space is really, really big. And the things in it are really far apart. Just to get to the nearest star, Proxima Centauri, you would have to travel 40 trillion kilometers. That's close to the approximate average distance between stars in our galaxy, which is 48 trillion kilometers. In a very real sense, we're stuck on a small island in a vast, almost unimaginably empty ocean.

But the problem with these long distances isn't that they're hard to cross. Space is mostly empty, so there isn't much in your way or any air dragging you down. The real problem is the *time* it takes you to cover those distances.

If you were to travel to Proxima Centauri at the fastest speed a human spaceship has ever gone (40,000 kilometers per hour), that journey would still take a long time—more than 100,000 years. Clearly, we need to go faster.

If you could manage to get your spaceship up to one-tenth of the speed of light (100 million kilometers per hour), you could reach Proxima Centauri in just over 40 years. That's a really long time for a vacation trip, but if you're planning to move there permanently, it might be worth it. And if you could go even faster, say one-half the speed of light, it would take less than 10 years.

But what about *beyond* Proxima Centauri? What if we wanted to visit other parts of the galaxy? The Milky Way is 1,000,000,000,000,000,000 kilometers wide, which means to get to the other side at half the speed of light would take about 200,000 years. Even if you could go at three-quarters of the speed of light, it would still take you 133,333 years to get there.

Fortunately, once you get up to three-quarters of the speed of light, physics helps you pass the time. At those speeds, relativistic effects start to become noticeable. Time passes differently for you when you're going that fast. From your point of view, the space in front of your ship gets contracted, so it feels to you like it takes less time to get there. If you get up to 99.999999 percent of the speed of light, the journey to the other side of the galaxy only takes 30 years from your perspective. Not bad!*

---

* Of course, by the time you get there, everyone you left behind on Earth will have been dead for tens of thousands of years.

The hard part, though, is getting your spaceship up to those incredible speeds. It takes a huge amount of energy. The formula for kinetic energy is roughly proportional to mv^2, where m is your mass and v is your velocity. That v^2 term is the kicker, because it means that doubling your speed requires quadrupling your energy. A medium-size ship carrying enough passengers and equipment to start a colony would probably weigh a few million kilograms, and accelerating that much mass to half the speed of light would require an absolutely *absurd* amount of energy: about five quadrillion megajoules, or the equivalent of one hundred times the energy consumed by everyone on Earth in a single year.

So where do you get this energy, and more importantly, how do you bring it with you?

The Toothpick Problem

One way to think about the problem of space travel is to think about the "toothpick problem." And by this we don't mean how to build a toothpick bridge between here and Proxima Centauri, but rather, "How do you accelerate a toothpick up to near the speed of light?" Now, this doesn't sound like a difficult problem. After all, a toothpick is pretty small, so how hard can it be? Well, it gets tricky when you consider *how* you're going to accelerate this toothpick out in space.

To infiniTEETH and beyond!

THE TOOTHPICK PROBLEM

The most common solution we have for pushing things in space is rockets. So you might think the answer is simple: just push the toothpick with a rocket. That's a big problem, though, because your rocket doesn't just have to push the toothpick; it also has to push all the fuel that you need to power the rocket. But the more fuel you bring, the heavier your space rocket gets, which means you need even *more* fuel. This cycle keeps going until, at some point, most of the fuel you bring is just to push the fuel itself. For example, to push a single toothpick up to about 10 percent of the speed of light, you need a rocket with a gas tank bigger than the planet Jupiter!

Of course, part of the problem is that rockets are really inefficient. They might be fun and exciting (plus they make the right kind of *whooshing* noise), but they are not a great way to get from one star to another. When you burn rocket fuel, you are breaking some of its chemical bonds, which releases energy. But that energy is a tiny fraction of the energy stored in the mass of the fuel itself. $E = mc^2$ tells you how much energy you could extract in principle from some fuel, and chemical burning only gives you about 0.0001 percent of that. To generate a joule of energy from rocket fuel, you need about a *million* joules' worth of mass.

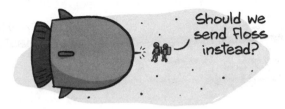

More Efficient Fuel

Can we do better than burning rocket fuel, which is basically nineteenth-century technology?

If we could find a fuel that is more efficient, then the toothpick problem gets easier. For example, if you can find a fuel that gives you more energy for the same weight, your toothpick wouldn't need such a large fuel tank.

But dealing with fuel that's more energetic is tricky, and potentially more dangerous. Below are some interesting options that might make space travel a lot easier.

Nukes

Nuclear power goes deeper than rocket fuel because it releases the energy stored inside the nucleus of the atom, not just the energy stored in the bonds between atoms. But we're not talking about building a nuclear reactor in your spaceship. That's still too inefficient. To make space travel work, we're talking about strapping nuclear *bombs* to the back of your ship and then setting them off. Nuclear bombs are much more efficient at releasing energy. If you build a ship where three-quarters of the mass is nuclear weapons, exploding them one by one can easily accelerate you to 10 percent of the speed of light.

This approach sounds promising, though there are a few hurdles. First, there's currently an international treaty that bans using nukes in space. Second, you would need *a lot* of nukes. To push a good-size spaceship packed for a long interstellar trip, you would need about two hundred times the number of nuclear weapons currently on Earth.

Ion Drives

If you aren't excited about riding a shockwave of nuclear blasts through space, there's a cleaner, more efficient option available: a particle accelerator (aka an "ion drive").

Usually, you build a particle accelerator to do science: you accelerate particles and then you see what happens to them when they smash into things. But you can also use them for space propulsion. Just like when you fire a bullet out of a gun, shooting particles gives you a small recoil. This is the conservation of momentum at work. If you create momentum out one side, you have to balance it with momentum in the other direction. Firing a bullet (or a particle) is like pushing someone away from you on a slippery lake: you'll both start moving.

An ion drive is just a big particle accelerator that fires particles out the back of the spaceship. It uses electrical energy to push the electrically charged particles and it's a very efficient way to turn that energy into speed. The drawback is that it gives a very gentle push, because the recoil you feel is as tiny as the particles. So you wouldn't be able to use an ion drive to take off from Earth's surface, but if you're out in space, it can push you for long enough to get you going to pretty high speeds.

pew pew pew pew!

I thought space travel would be more dignified.

The tricky part of the ion drive is getting the electrical energy. To get enough of it for a long space trip, you'd need a heavy fusion reactor or giant solar panels, which is going to add to your mass and bring down your efficiency. Fortunately, particle physics offers a potential answer to this problem as well.

Antimatter

To power an ion drive, we'd like a source of energy that is as efficient as possible, and there's nothing more efficient than something that converts *all* of its mass into energy. Such is the power of antimatter.

Antimatter is a real thing, not science fiction fantasy. Every kind of matter particle we have discovered has a corresponding antiparticle. Electrons have antielectrons, quarks have antiquarks, and protons have antiprotons.* There are big mysteries as to why anti-matter even exists, but the important thing is what happens when matter and antimatter meet.

WHEN YOUR SPACE OPERA
TURNS INTO A SOAP OPERA

When antimatter bumps into the regular kind of matter, they both annihilate, converting all of their mass into energy. For example, if an electron meets an antielectron, they turn into a photon, a particle of light. The same is true of all matter-antimatter pairs. It's so efficient that a small quantity of antimatter can combine with a small quantity of matter to release a lot of energy. If a raisin were to bump into an antiraisin (a raisin made of antiparticles), it would release more energy than a nuclear explosion.

While this idea sounds promising, it's also a very dangerous one. If any of your antimatter fuel were to touch your (regular matter) ship: *boom*. Generally speaking, you want a controlled release of energy to power your ship, not a sudden explosion that tears you to shreds. Containing antimatter is *hard*. You might imagine using magnetic fields to contain it, but they might not work for very long. One tiny leak and it's anti-adios.

Another problem with antimatter fuel is figuring out where to

---

* We're not sure yet if neutrinos have separate antineutrinos or if they are their own antiparticle.

get it. While we currently have the technology to create it in high-energy particle collisions, it's astonishingly expensive. The collider at CERN makes picograms of antimatter every year, at a cost of several hundred trillion dollars per gram. Scaling up production enough to power a whole spaceship might be cost-prohibitive.

Black Hole Power

Another possible idea for powering your spaceship that is 100 percent efficient is to use a black hole. Black holes are the most compact way to store energy in the universe.

As it turns out, they also emit energy. Black holes generate something called "Hawking radiation," which scientists predict happens when a pair of particles is created near their edge. This creation of particles occurs all the time in regular space due to quantum fluctuations. But when it takes place at the edge of a black hole, something interesting can happen. The particles get a little boost of energy from the black hole's gravity, essentially borrowing some of its energy. If one of the particles escapes while the other one gets sucked back in, then the one that escaped carries away some of the energy of the black hole. For a black hole, losing energy basically means it loses some of its mass. In this way, the black hole essentially converts some of its energy into radiation, shooting off particles from just outside its edge. If you could capture these particles, you could use them to power your ship.

For a large black hole, this radiation is very faint, but physicists think that for smaller black holes it's much more intense. A "little" black hole that weighs as much as a couple of Empire State Buildings would give off a lot of particles and be very bright, gradually turning all of the energy stored in its mass into radiation.

The idea would be to put that black hole in the center of your ship, and build your ship so that it deflects all of that radiation out the back. That impulse would be enough to push your ship forward. And when your ship moves forward, its gravity then pulls the black hole behind it, keeping your crazy black hole contraption together.

Making small black holes for fuel wouldn't be easy, but if we can do it, scientists think that they would last a few years, pumping out energy until they evaporate into nothing.

Sailing Away

If the thought of riding a spaceship powered by nuclear explosions, deadly antimatter, or dangerous black holes makes you rethink this whole idea of visiting other stars, we totally understand.

Unfortunately, if you are stuck on the idea of packing all the fuel you'll need for your trip, then it's hard to find a more efficient fuel than those three.

But what if there was a different way of navigating the vast oceans of space? What if you could literally *sail* your way to another star or planet?

That is, after all, how humans first navigated the high seas. We didn't bring all of our fuel like we do now. Sailors relied on the wind to push them to their destinations. What if something like that was available for space travel?

Solar sails sound a little silly, but they are a real option, and a proven technology. The idea is for your ship to have a large, wide area that can catch particles, just like a sail catches the wind. When the particles bounce off the sail, they transfer their momentum to it, giving your ship a push.

Where would these particles come from? Fortunately, we have a massive energy source that generates high-speed particles: the Sun. It's an awesome ball of fusion that constantly shoots photons and other particles out in every direction. All you'd have to do to sail your way out of the solar system is to point your particle catcher toward the Sun and let its rays and radiation gently push you toward the cosmos.

One caveat is that the Sun's rays aren't enough to get a ship up to the high speeds needed for rapid interstellar travel. The solar wind gets pretty faint as you get farther away from the Sun. A potential solution to this is to build a giant laser here on Earth and aim it at your departing spacecraft, essentially pushing it from here. Another solution is to build enormous mirrors that focus the Sun's energy. Both of these ideas could give your ship the acceleration it needs to get to a tenth of the speed of light or more.

So What Are We Waiting For?

We get it. Some of the ideas we discussed here seem a bit wild. But from a physics perspective, they are all technically feasible! That means there's really nothing stopping us from visiting other stars. We know how to do it; we just have to . . . do it. It might be expensive and complicated, but the physics is not the problem. It's almost like the universe is daring us to do it. Does it seem impossible to create and wrangle a black hole? Or bottle antimatter without touching it? Sure! But think about all the things humans have been able to achieve that we once thought were impossible.

All we need is the vision to imagine it and the will to see it through. We are called by the cosmos to set our sights on that farthest of horizons. Let's fire up our inner explorers and look up . . . to the stars!

Is an Asteroid Going to Hit Earth and Kill Us All?

You never see it coming.

That's the conventional wisdom about how people usually meet their doom. Life is full of surprises, including how it ends.

This might be particularly true for us as a human species. After all, space is a dangerous place, and we are squishy beings desperately hanging on to a small planet careening around in the dark. Out there is a vast, unknowable void filled with exploding stars, supermassive black holes, and potentially nefarious alien beings.

Fortunately, as far as we know, supernovas and black holes (and aliens) aren't happening anywhere near us anytime soon, but there *is* the danger that something might come for us and spell our early demise: rocks. Space is full of huge rocks that are constantly whizzing around at enormously high speeds, smashing into anything that gets in their way.

If there's any doubt that space rocks can be dangerous, just look at the surface of any moon or planet in the solar system that doesn't

have a protective atmosphere. You'll see zillions of craters, some of them thousands of kilometers wide, and each one evidence of a violent cosmic collision. Our own moon, for example, has millions of craters, which is more craters than a teenager has pimples.

This makes a lot of people wonder: Are we next? What are the chances that a big rock could smash into Earth and kill us all? And where are all these speedy rocks coming from anyway?

Rocks in Space

When you think of dangerous, giant asteroids, you probably assume they come from the depths of space, far beyond our solar system. But actually, a killer rock is most likely to come from right here in our own backyard. That's because interstellar space is fairly empty, whereas our solar system is *full* of big, deadly rocks. What follows is a tour of the main clusters of space rocks in our neighborhood.

The Asteroid Belt

The first group of space rocks is the asteroid belt, a collection of rocks between Mars and Jupiter. There are millions of rocks in the asteroid belt. Most of them are small, but hundreds of them are more than 100 kilometers wide, and some are as large as 950 kilometers (which is about the size of Montana). If any of those larger rocks were to hit Earth, they would probably kill us all.

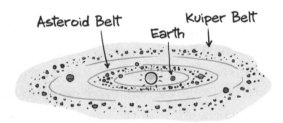

The Kuiper Belt

The second-largest concentration of asteroids nearby is the Kuiper Belt, a big disk of ice balls past Neptune. The Kuiper Belt contains around *one hundred thousand* ice rocks that are more than 80 kilometers in diameter, and these are still pretty dangerous.

The Oort Cloud

Finally, there's the Oort cloud, a vast cloud of ice and dust far beyond Pluto, where most of the comets that we see come from.

Astronomers theorize that the Oort cloud has trillions of icy space rocks bigger than a kilometer in size, and billions more that are larger than 20 kilometers.

It turns out that our cosmic neighborhood is not quite the neat and tidy place you thought it was. It's actually full of litter!

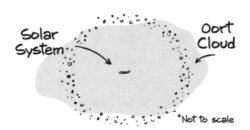

How did our space neighborhood get so full of rocks? It all goes back to the beginning. Our solar system was made out of gas and dust and little pebbles. Some of these materials were created during the Big Bang, and others are the remains of stars that burned themselves out and exploded. Most of the gas, which is lighter, pulled together in the center into a blob so dense that the gravitational force ignited it into a star, making our Sun. A lot of the rest clumped together in the periphery, and since there wasn't enough gravity to ignite them into stars, they became planets with hot, molten cores due to gravity's pressure. But not all of the remaining bits of rubble were swept into the Sun or planets. The large amounts of it that were left over grouped together into smaller bits that are still zooming around the solar system.

At first, the solar system was a chaotic place. Everything was new, and all these young planets and chunks of rocks were fighting to settle into their orbits. You might think that you had a nice planet going, when all of a sudden . . . *boom*, you would crash into another giant rock that had the same idea. Scientists think that's how our moon was formed: a big asteroid smashed into the new-born Earth, blowing a big chunk of it out into near orbit.

Ooh, that's gotta hurt!

Fortunately, the solar system is an old place now, and the early wild days of bouncing and bashing have quieted down. By now most things in the solar system are in stable orbits. Anything that wasn't probably crashed by now, or learned to go with the flow of the rest of the planets and asteroids. It's like one of those crazy roundabouts that you see in Europe where everybody is speeding around extremely close to one another. They've been doing it for years, so you can be pretty sure they know what they are doing.

%.$#^@!*   #@*~&!*

*Translated from Italian

But that doesn't mean that we're out of danger. Some of those asteroids or ice balls might still be on a trajectory that can impact Earth in the future. Or they might switch to a trajectory that crosses our path. Sometimes these rocks can get knocked out of their own orbits and can cause trouble. For example, the distant Sun can slightly warm one side of an asteroid, which shifts its orbit. It can crash into another rock, which then collides with another rock, and so on. And if any of them interact with Jupiter's gravity, they can be pulled into the inner solar system. Before you know it, you might have a thousand-rock pileup on the Inner Solar System Highway, and you're doing insurance paperwork for a billion years.

How Bad Would It Be, Really?

What would happen if an asteroid hit Earth? It depends.

Before a rock actually hits Earth, it has to go through the atmosphere, which gives us some protection. The particles of air drag on the incoming rock and slow it down, like a cushion that absorbs its impact. Think of a bullet fired into a pool of water, or a bowling ball dropped into a huge vat of Jell-O.* The air particles can't get out of the way fast enough, and the energy of the space rock compresses them into a shock wave. When air—or anything—gets compressed, it gets hot. In this case, the temperature at the shock front

I'm coming in hot!

---

* Really, think about that. It's a fun mental image.

can get up to 3000°F. This is why our space shuttles and landing modules heat up on reentry from orbit, and it's why we put advanced ceramics and cooling systems in front of them to deflect and absorb the heat created by this air drag.

Rocks from space don't usually come with fancy shields to keep them cool, so they just get hot. *Very* hot. Depending on how hot they get, they might shatter in the atmosphere, exploding into smaller fragments that rain debris on the surface, or they might hang together and deliver most of their energy directly to the surface of Earth.

Small rocks (up to about one meter wide) actually hit Earth all the time but burn up in the atmosphere as shooting stars. If you catch them on a clear night, they are even beautiful to watch.

But as the rocks get bigger, they start to get more dangerous, and not even our atmosphere can stop them. To get a sense of scale, the table on the next page compares the amount of energy that asteroids of different sizes would have with the explosive power of the bomb dropped on Hiroshima in World War II.

Rocks that are five meters wide have about the same energy as the bomb dropped on Hiroshima. This sounds bad, but actually, scientists aren't too worried about them. These rocks often strike somewhere in the ocean or explode in the very upper atmosphere, usually far from populated areas.

Size of Asteroid	Explosive Power
5m	1 Hiroshima Bomb
20m	30 Hiroshima Bombs
100m	3,000 Hiroshima Bombs
1 km	3,000,000 Hiroshimas
5 km	100,000,000 Hiroshimas

Stepping up to twenty meters in size (about five elephants wide) means a rock carrying the same energy as *thirty* Hiroshima bombs. That's a huge explosion. If we get really unlucky and a rock that size makes it through the atmosphere and hits somewhere like Manhattan, it would be an enormous disaster. Millions of lives would be lost. But it wouldn't necessarily mean the end of the human race. In fact, a twenty-meter asteroid exploded in our atmosphere very recently.

In 2013 over Chelyabinsk, Russia, a twenty-meter-wide rock from the asteroid belt hit our atmosphere at sixty thousand kilometers per hour. It was midmorning, but the light from the explosion was reportedly brighter than the Sun and could be seen up to one hundred kilometers away. Around a thousand people were injured. It was spectacular enough to cause panic and widespread religious reawakenings, but not spectacular enough to end humanity's time on Earth.

Above that size (in the kilometer range) is where the danger zone for our species really begins. Scientists believe that the last arrival of a multiple-kilometer-size rock was sixty-five million years ago, and that it might have been what caused the extinction of the dinosaurs.*

You might be asking yourself: If Earth is thousands of kilometers wide (12,742 kilometers, to be precise), how can a relatively small rock of a few kilometers cause so much destruction? Let's consider the case of a humble, five-kilometer rock.

A five-kilometer-wide rock falling to Earth would carry something on the order of 10^{23} joules of energy. For comparison, the average American uses about 3×10^{11} joules of energy in a year, and all of humanity uses about 4×10^{20} joules. So this one collision would carry a thousand years' worth of human energy, all concentrated and delivered rapidly in a single spot. In nuclear-weapon units, that's two billion kilotons, or about one hundred million times the energy in the Hiroshima bomb.

That much energy released on land would create an explosive shock wave that would travel rapidly from the impact site, carrying with it enough heat and wind to destroy anything within thousands of miles. It would also cause earthquakes that shatter all the

* Interestingly, scientists think the rock that killed the dinosaurs (about ten kilometers wide) flew by Earth years before it actually hit our planet, which should have given the dino-scientists some warning.

land around it, and trigger enough volcanoes to soak the entire area in hot lava.

If you are anywhere near that impact, your fate is simple: you are toast. Charred, blackened toast that can't be improved no matter how much you butter it. How close do you have to be? In this scenario, Los Angeles is probably not far enough away from an impact that happens in New York.

But even if you are far from the impact area (say, on the other side of the world), you probably won't survive for very long. You might avoid the immediate blast, but you would still suffer the earthquakes and reignited volcanoes that the impact causes. Your larger problem, though, will be the cloud of superheated dust, ash, and rock fragments that will be thrown up into the atmosphere. Some of this superhot dust will drift away, roasting Earth's surface and torching forests. And it will hang out in the skies for a long, *long* time. This cloud would shroud Earth in darkness for years, or decades, or longer, which is probably what killed the dinosaurs.

You might wonder what would happen if the asteroid hits water rather than land. Unfortunately, things don't get much better. First, a lot of the initial energy would be absorbed by the water, creating a mega-tsunami with waves *several kilometers* high. Imagine yourself looking up at a wave that's four or five times bigger than the

Empire State Building. A wave that big means that *Denver* would suddenly become beachfront property, and Australia and Japan would get completely wiped off the map.

And that's just the immediate aftermath. A giant cloud of dust would probably kill most of our ecosystem, making life as we know it pretty unsustainable. And if the asteroid hits water, the impact would also put enough water vapor in the atmosphere to cause an accelerated greenhouse effect. This greenhouse effect would trap energy on Earth and heat the planet up to unlivable temperatures.

That's just what a five-kilometer rock would do. Now imagine what an even *bigger* asteroid would do!

How Likely Is It?

To get a sense of how likely it is for a large asteroid to hit us, and whether we would see it coming, we spoke to the good people working for NASA's Center for Near-Earth Object Studies (CNEOS), headquartered at the Jet Propulsion Laboratory in Pasadena, California. Really, their name should be "Asteroid Defense Force" because they are the group tasked with preventing the complete annihilation of the human race by a giant rock crashing into us. (And you thought *your* job was important.)

The main approach of CNEOS (together with their international collaborators) is to look for and keep track of all the rocks in the solar system, so that we might have a heads-up if any of them are on a path to hit us. Using telescopes, and after decades of hard work, the CNEOS team has created a pretty good database of all the biggest rocks around us, where they are, and where they will be in the near and far future.

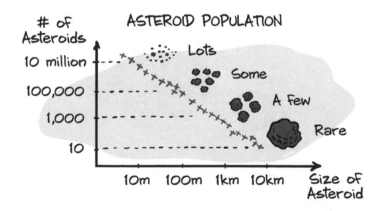

What they've found is that there is an inverse correlation between the size of the rock and how many of them there are in the solar system. Small rocks are plentiful in our neighborhood, but the really large rocks are hard to find. In other words, the bigger the rock, the rarer it is. This is good news because the rarer a type of rock is, the less likely it is to crash into us.

For example, CNEOS estimates there are hundreds of millions of rocks out there that are around one meter in size. That's a lot of rocks, and in fact, rocks of this size hit Earth all the time, about five hundred times per year. That means that on any given day, there is probably one of these rocks crashing somewhere on Earth. Fortunately, they cause very little damage.

OW, OW, OW...

As the rocks get bigger, they get rarer. For example, rocks that are five meters wide are in the tens of millions in the solar system, and strike Earth only about once every five years. Twenty-meter-size rocks (like the one that exploded over Chelyabinsk, Russia) are in the single millions, and on average only hit Earth every fifty years or so.

But what about the really large ones? Even if those are rarer (there are only a thousand rocks that are one kilometer wide, and just a few dozen greater than ten ki-lometers), you only need one of them to hit us to potentially end the human race.

Luckily, large rocks like that are not only rare but also relatively visible. If a large rock is on a regular orbit, it's likely that we'll have seen it reflect light from the Sun. That means the CNEOS team is fairly confident they know where most of them are. They've counted them and mapped their trajectories, and so far, none of them seem to be on a collision course with us.

At least, we don't think so. The good news is that we know where 90 percent of the big rocks in the solar system are. The bad news is that we don't know where 10 percent of the big rocks in the solar system are.

There could still be large rocks out there that we haven't seen. They might be hidden, or they could be on an orbit that hasn't brought them close enough for us to observe them. Remember that asteroids don't glow on their own, and a few kilometers in size isn't that big compared to the size of our solar system. That means there's still the possibility that a large asteroid can sneak up on us out of the darkness of space.

Deadly Snowballs

Much more concerning to the scientists at CNEOS are the other type of space rock that can hit us: giant snowballs (aka comets). While NASA has a good handle on most of the asteroids in the solar system that can kill us, it turns out that comets are much harder to spot.

Most comets we see are huge balls of rock and ice that fall in from the Oort cloud toward the Sun on very long orbits. Sometimes these orbits can take hundreds or thousands of years to go around the Sun. That means that when a comet visits the inner solar system (our cosmic neighborhood), it might be the first time that we see it.

Even worse, after their long trip from the cold cosmic suburbs, they are going to be moving much more quickly than an asteroid, which means (a) we won't have time to react (a year at best), and (b) they would have a more devastating impact if they hit us.

Scientists think that the chance of a comet crashing into us is probably rare, but it's hard to estimate. Very recently, it happened to one of our neighbors: in 1994, the comet Shoemaker-Levy 9 broke into twenty-one pieces on its way toward the sun, and those fragments crashed into Jupiter. One of those pieces created a gigantic explosion approximately the size of Earth.

In fact, it was this comet collision that sparked NASA to create the Near-Earth Object program to catalog and track all near-Earth objects. After all, if it happened once, it can happen again, and maybe to us.

What Can We Do About It?

Let's say a comet suddenly comes out of the blue and is on a path to hit us. Or let's say that we find a new big asteroid we haven't seen before and we learn that its orbit intersects with ours in the future. Or let's say that some kind of solar system event knocks a big rock straight toward us. Is there anything we could do about it?

In the movies, all it takes is a musical montage of scientists in lab coats, a pot of coffee, and a whiteboard full of scribbles to figure out a solution (plus, it helps if you have Bruce Willis). But is that realistic?

Surprisingly, this is something that groups like CNEOS actively think about, and according to them, the strategies for surviving a big rock coming our way fall into one of two categories.

Option #1: Deflect

The first option is to try to deflect the asteroid or comet—that is, to nudge its trajectory so it's not on a collision course with us. Scientists have a few good ideas for how to do this:

Rockets: This plan involves shooting a rocket at the incoming rock to either crash into it or blow up enough of it to change its trajectory. It might also be possible (though less likely) to land on the rock and use the booster engine to push the rock into a new trajectory.

Digger: Another idea is to send a giant crane or robot to land on the rock and start digging, pushing the debris out into space. The momentum from all the debris would essentially cause the rock to change course.

Lasers: Another fun idea would be to build a huge laser here on Earth and then shoot it at the asteroid or comet. The goal would be to heat up one side of the rock so that the melting ice or vaporized rock would push the rock out of Earth's path.

ASTEROID DEFLECTION TECHNIQUES

Mirrors: If you wanted to get really fancy, you could send a set of lenses and mirrors to gather sunlight and focus it onto the rock. This would boil off some of its material, pushing the rock out of its collision course.

Option #2: Destroy

The second option is, of course, to try to destroy the big rock before it gets to us. In other words, nuke it.

One idea is to launch a nuclear missile that intercepts the rock and blows it up, hopefully shattering it into smaller pieces that would then burn up in our atmosphere. Some of them may still hit the ground, but that's a better outcome than if the whole rock struck Earth.

On the other hand, it might be that the incoming asteroid is mostly just a pile of rubble loosely bound together by gravity. In that case, a single nuclear explosion wouldn't be very effective at dispersing the rock, and we would be better off sending a series of smaller nukes. Perhaps we would optimize the distance of the nuclear blasts for maximum dispersal, perhaps setting them off a bit above the surface so that they do more deflecting than destroying.

Of course, the most crucial factor that determines whether any of these strategies will work is how much time we have. According to

Good thing we have extra nukes.

CNEOS, "the three most important things you need in order to survive an asteroid or comet impact are: (1) early detection, and (2) the other two don't really matter much."*

If we have a lot of warning (years, hopefully), then we might have time to build and deploy one of these strategies. Not only that, but more time also gives us greater opportunity to affect the outcome.

For example, if we learned that a particular asteroid would strike Earth in one hundred years, any small nudge we give it today would have a huge impact in its future trajectory. It's like firing a sniper rifle at a target one kilometer away. The slightest deflection of the rifle to the side will give you a large sideways displacement of the bullet over the full kilometer it travels. The same is true for asteroids: if you see one coming far enough in advance, you only need to give it a little push to knock it off course.

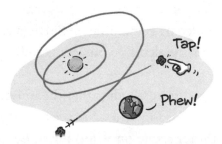

This is why it's so important to keep track of all the asteroids and comets flying around us, and why the idea of one of them coming out of the blue is so scary.

---

* That's an actual quote from Dr. Steve Chesley, senior research scientist at CNEOS, who graciously agreed to be interviewed for this chapter.

Should You Worry?

Before you start building that underground bunker or go on a shopping spree for canned foods, we should probably tell you that the probability of an asteroid coming to kill us all is actually not that high.

In the short term, the team at NASA and the few dozen people working on this around the world are doing everything they can to spot these rocks early, and they are quietly doing a solid job so that you don't have to keep looking up with anxiety. They are confident that nearly all of the planet-killing rocks have been spotted and are accounted for and pose negligible risk to Earth. There are even plans for more powerful telescopes, like the Near-Earth Object Surveillance Mission space telescope and the Vera Rubin ground telescope, that will vastly increase humanity's ability to detect rocks earlier. Thinking about your personal risk, you are much more likely to get killed by something on Earth (a car crash, a fall in the shower, getting strangled by your pet gerbil) than a rock from space.

But it's always good to remember that the universe is unpredictable, and that there are limits to our science. Perhaps there is a large asteroid with our name on it lurking in our solar system, or maybe there is a comet coming from afar aimed straight at us. Predicting anything with certainty in a complex solar system like ours is a tough order. Do you remember that asteroid that exploded over Chelyabinsk, Russia? It came out of nowhere. The only warning we had was when it hit the atmosphere.

The truth is that we live in a chaotic cloud of rocks and planets, all of them pushing and pulling on one another in an intricate gravitational dance. Every collision or close approach should give us pause, and it should motivate us to support science further so we can understand more about our galactic neighborhood. It should also get us thinking about humanity's ability to work together, and whether we can set aside our differences for our own survival.

Because if we can't dance that dance, well . . . just remember what happened to the dinosaurs.

Didn't see it coming.

Are Humans Predictable?

L et's take a moment to think about the choices you make. For example, you chose to pick up this book, and you're choosing right now to read these words. There, you did it again. You made the choice to read *these* words, too. *And these.*

AND THESE.

Okay, by now you're probably tempted to *stop* reading these words, just to prove that you have a choice and that we are not in control of you. After all, you have free will, right? Go ahead and look away for a moment if it makes you feel better. We'll wait for you.

Are you back? Great choice. (We totally predicted that.)

The point is that we all like to think that we are in charge of our own actions. As we go about our day, we make hundreds, if not thousands, of decisions. Should I get out of bed or hit the snooze button? Should I shower today? Should I eat bacon and eggs for breakfast, or a warm bowl of oatmeal? The world is your oyster, and if you wanted to have oysters for breakfast, you could do that, too. We don't recommend it, but hey, it's your choice.

It's your choice.

This feeling of control makes us uncomfortable with the suggestion that any of our choices are predetermined, or that they can be predicted. We like to believe that when we decide to do something, it happens *at that moment*, not before, and that no one could have foreseen it.

But is that really true? Are our choices truly unpredictable? As science has advanced and our understanding of the laws of physics has grown more and more complete, a lot of people have begun to wonder if it's possible to predict what a person is going to decide. Or to take the question out of the lab and into the halls of philosophy: Do we really have a choice when making decisions? Or can the actions of a complex, thinking being be reduced to a simple set of predictable laws?

The answer, should you choose to read it, lies ahead. But be warned: we predict you probably won't like it.

Physics in Your Brain

As far as we know, everything in the universe follows the laws of physics. To date, we haven't found a single thing that doesn't follow them. The laws we've discovered and have improved over the centuries seem to be true for everything, from bacteria to butterflies to black holes.

And since *you* are also in the universe, the laws of physics apply to you, too, and to your brain, which is the center of your thinking being. Brains and black holes are made of the same stuff (matter and energy), so the same rules that apply to black holes also apply to brains.

How can physics help us understand the brain? Is there a law that predicts how many cookies you will eat today, or whether you'll choose a banana instead? Unfortunately, there's no outright Newton's second law of cookies, or Einstein's banana brain equation. Physics can describe something like the brain, instead, by breaking it down into smaller, simpler pieces that we can understand. Then we add all the pieces together to see how the whole thing works.

This is just like when you were a kid and you took a toaster apart to see how it worked. Hopefully, unlike your toaster, we'll be able to put your brain back together again.

Your brain can be broken down into lobes, and those lobes can be broken down into neurons. Each neuron is essentially a little electrical switch that gets "on" or "off" signals from other neurons. Then, based on those signals, the neuron might send an on or off signal out to other neurons.

Your entire brain is made up of these neurons. A tangle of eighty-six billion of them, all wired together by over one hundred trillion connections. Together, this huge network of simple biological switches makes up who you are: your memories, abilities, reflexes, and thoughts.

And that's pretty much it. That's all your brain is: a bunch of simple switches, and lots of connections.

Just like an electrical switch, each neuron's output is determined by the input it gets and the little biological circuit inside of it. Neurons don't have moods or whims. They don't fire because they "feel" like it. Each neuron simply follows the rules that are programmed into its genetic makeup.*

---

* It's a little more complicated than that, because neurons can change and adapt. But even then, each neuron is following rules for how to change and how to adapt, so the point is the same.

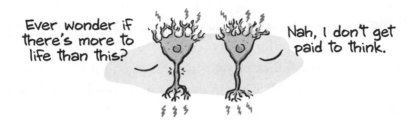

Does that mean that brains are predictable? After all, if a neuron is simply following rules, then you should be able to predict what a neuron will do. And if you can predict what a neuron will do, then you should be able to predict what a whole bunch of neurons connected together will do. And if you can do *that*, then you could in theory predict what a human will do.

Not so fast. There are certain things about the brain that make it not so easy to predict. Those things have to do with chaos theory and quantum physics.

A Chaotic Brain

While neurons don't have moods, they are still a sensitive bunch.

Even if something is purely mechanical, like a perfectly tuned machine or a rigid computer program, that doesn't mean it will always give you the same result. For example, you don't always get heads when you flip a coin. Even though a coin follows the laws of physics when you throw it up in the air and when it hits a surface, it's still very hard to get it to land on the same side every time. That's because a coin flip is very sensitive to small changes in how you throw it: a small flick of your fingers, an errant air current, or a tiny bump on the table where it lands can affect whether the coin settles on one side or the other.

In the same way, neurons are very sensitive to small changes in their inputs. Neurons work by adding up the on or off signals they get from other neurons, weighing them depending on the strength of each connection. If the sum of all the signals crosses over a threshold, then the neuron activates and sends an on signal downstream to all the neurons connected to its output. But if the sum doesn't meet the threshold, the neuron stays silent. You can imag-

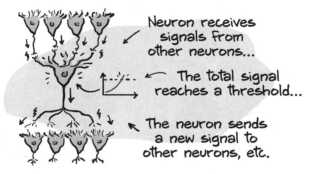

THE NEURON GAME

Neuron receives signals from other neurons...

The total signal reaches a threshold...

The neuron sends a new signal to other neurons, etc.

ine that a single input signal (out of thousands), or one small change in the strength of one connection, can make the difference in whether a neuron gets activated.

This sensitivity gets even more dramatic when you connect a lot of neurons together. A small change in one neuron can lead to a cascade of consequences that give the network a whole different output. For example, that small change can mean the difference between whether you choose to eat the cookie or the banana.

When a system is that sensitive to small changes, physicists say that it's "chaotic." It's the same reason why physics can't predict the weather very well. We can predict what one raindrop does, but weather is made up of a lot of water drops and air molecules that are sensitive to being bumped by each other (and by the wind, the mountains, cold air pockets, etc.). These effects don't cancel each other out; they build on each other, becoming larger and larger. If you have zillions of drops, then getting the direction of one of them slightly wrong now might make your prediction about tomorrow's rainstorms totally wrong. Add pesky butterflies that keep flapping their wings and the whole thing becomes so chaotic that it's impossible to predict.

Like rainstorms, brains are chaotic. You might try to predict the behavior of one neuron and do a pretty good job. But what happens if your prediction isn't perfect? For example, maybe your model of one neuron is 99 percent accurate, which is pretty good. (A 99 percent on a math test is an A+.) But 99 percent right also means 1 percent wrong, and when you try to predict the next set of neurons, those errors are going to propagate and grow. Extrapolate that to eighty-six billion neurons and you can see why predicting what your brain is going to do is very, very hard.

Eventually, though, science might figure out how to predict even the weather. If you have enough computing power (and enough time), it is theoretically possible to simulate anything to a perfect degree. In fact, the majority of the world's supercomputers today are dedicated to building increasingly more precise models of Earth's weather. You might imagine a computer in the future so

large and powerful that it can perfectly simulate every neuron and connection in your brain with accuracy down to the molecular level.

Does that mean that in the future, scientists might be able to create some new kind of supercomputer to model your brain and predict your snacking decisions? Not if your brain is also quantum mechanical.

Your Quantum Brain

If your brain is chaotic, does that mean that it's unpredictable? Not necessarily. Just because a system is chaotic doesn't mean that it's unpredictable. It can be hard to predict what it's going to do, but it's still predictable. After all, it's still following the laws of physics, and the laws of physics can be simulated and therefore predicted.

But what if the law of physics themselves can make something unpredictable?

When you peel back a few layers of reality and look at the particles that make up everything around us, you learn something strange about the universe: the rules that apply to perfect machines and computer programs don't also apply to quantum particles.

Ideally, giving a system the same input conditions will give you the same output, but this *is not true* for quantum particles like electrons. What does that mean? It means that if you poke a quantum particle exactly the same way more than once, it *doesn't* always respond the same way. One time it might bounce back; another time it might ignore you entirely.

Hey, watch where you're poking!

Whatever.

QUANTUM FICKLENESS

How is that possible? Well, electrons still follow the laws of physics, but they do it in a particular way. The laws of quantum physics don't specify exactly what happens to an individual electron. Instead, they specify what is more or less *likely* to happen. What *actually* happens for a single individual electron is randomly drawn from that list of possibilities. In other words, the laws of physics at the quantum level don't tell you what *will* happen; they tell you what *can* happen, and with what odds.

Poke the same electron the same way several times, and you will get different outcomes each time.* If you poke it enough times, you'll start to see a pattern (e.g., that it bounces back 75 percent of the time, and ignores you the other 25 percent of the time). This pattern *is* predictable by the laws of physics. But on any particular

---

* Doing the same thing over and over again and expecting different results is sometimes used as a definition of insanity. But it's totally reasonable in the quantum realm!

poke, what the electron does is *not* determined by the laws of physics but by a completely random choice made by the universe (not by the electron).

If that seems bonkers, it's because it *is* bonkers. We're used to things having a clear cause and effect: if I push on a chair, the chair will move in that direction. But that only happens at the macroscopic level. At the microscopic level, things are truly random.

This is important to our question because neurons are made out of quantum particles. In fact, everything you know is made out of quantum particles, which are not predictable.

Wait, What?

At this point, you might be a little confused. We just told you that neurons are made out of quantum particles, and that quantum particles are random (and thus unpredictable). Does that mean that neurons are also unpredictable?

Again, not necessarily.

Looking around, we don't notice a lot of weird quantum effects. We don't see cookies randomly disappearing from the package, just popping out of existence or quantum tunneling into your stomach.

Cookies and other large things seem to follow predictable rules. So why are large things so different from small things?

There are two reasons for that difference: (a) the randomness of quantum particles is very, very small compared to your cookie, and (b) for most things in our world, that randomness averages out to nothing. Let's tackle these ideas one by one.

The Randomness of Quantum Particles Is Very Small

Quantum particles are extremely small compared to a cookie or a neuron. A single neuron is made of more than 10^{27} particles. So the quantum fluctuations of a single particle (whether it moves here or there) are so small that they aren't likely to make a big difference. For example, if a single cell in your body moved to the right a little bit, would you feel it? Probably not.

The Quantum Randomness Tends to Average Out

More likely than not, the quantum fluctuations of all the particles in a neuron will cancel out. If the neuron has a particle that does a weird quantum move to the right, that effect will most likely be corrected by another random particle that moves to the left. In other words, any little unpredictable quantum wiggle tends to get drowned out by the wiggles from all the other particles.

These two ideas are true of everything that is much bigger than quantum particles. In fact, that's why it took physicists a long time to discover quantum mechanics: because you only see it on really, really small things. If basketballs and raindrops were suddenly veering off course or acting randomly, we would have discovered quantum physics much earlier.

But remember, just because quantum effects are small and usually average out doesn't mean that they can be totally ignored. Are big things, like neurons, totally unaffected by quantum randomness? The truth is, we don't know! It's conceivable that neurons *are* sensitive to random quantum fluctuations in a significant way, and that these fluctuations affect whether a neuron gets activated. If that's the case, then there *would* be an element of randomness in our brain circuits, which means you could never truly predict what anyone is going to think or do.

Unfortunately, we don't have any evidence right now that neurons are sensitive to quantum randomness. Some famous physicists have supported this idea, but so far there haven't been any experiments that show that neurons exhibit true quantum randomness. Other physicists try to make connections between quantum randomness and philosophical concepts like consciousness and free will. But so far, these arguments are as persuasive as Nigerian email scams.

Did We Get It Right?

To summarize, your brain is both chaotic *and* quantum. But to what extent that means you are predictable is debatable.

If the brain is sensitive to quantum mechanical effects, then your decisions have a random element that is impossible to predict. Not just hard, *impossible*. As in, literally nobody knows what you are going to do next.

And even if your brain is not sensitive to quantum mechanical effects, chaos theory makes it almost impossible for anyone, or anything, to predict what you're going to think and do. While it might be possible in principle to perfectly simulate your eighty-six billion neurons and their one hundred trillion connections, it's almost certainly impossible in practice in the near future.

So it seems that, for now, you can rest assured that your brain (and therefore you) are not predictable. But is that the same thing as being in control of your decisions?

Being unpredictable is not quite the same thing as being *in charge*. Randomness is not the same thing as having control. If your brain is *random*, that doesn't mean that *you* are making any decisions; it means the universe is tossing dice and deciding what you do. Maybe your "you-ness" is the same as everyone else's you-ness: you (and we) are the universe.

If you rolled your eyes at that new age conclusion, then we can totally predict what you're going to do next: you're going to stop reading this chapter.

Where Did the
Universe Come From?

When you look up at the majesty of the night sky, or marvel at the intricate beauty of the microscopic world, you can't help but wonder: Where did it all come from? Why does the universe exist at all? What, or who, is responsible for this?

People have been marveling at the universe and guessing at its origins for a long time. Certainly, for far longer than we've had physics or cartoons. These questions are important because they might shed light on the context of our existence. We want to know *how* we came to be because it might tell us *why* we are here and how we should spend our time. If you knew where the universe came from, it might change how you live your life.

So what can physics actually tell us about this, the biggest of all questions?

In the Beginning

Before we ask where the universe came from or how it came into being, we need to back up a little. The first question we should ask is: Did the universe come into existence, or was it always here?

You might be surprised to learn that physics has a lot to say about this question. Unfortunately, a lot of what it has to say isn't very unified. In fact, our two great theories about the universe, quantum mechanics and relativity, point us in two very different directions on this topic.

The Universe came from ... that away!

The Quantum Universe

Quantum mechanics tells us that the universe follows unfamiliar rules. According to the theory, particles and energy behave in strange and uncertain ways. That can be pretty confusing, but fortunately, that's not the part of quantum mechanics that's relevant to the question at hand. That's because quantum mechanics is actually *crystal clear* about the past and the future of the universe.

Quantum mechanics describes things in terms of quantum states. Quantum states tell you the probability of what might happen when you interact with a quantum object. For example, it might tell you the probability of a particle's position. You might not know where a particle is right now, but you can know where it's *likely to be.* Quantum states are interesting because if you know what the state is for a quantum object today, you can use it to predict what the state is going to be tomorrow. Or in two weeks. Or in a billion years. The most famous equation in quantum mechanics, the Schrödinger equation, isn't about cats and boxes. It tells you how to take what you know about the universe and project it forward in time. And it also works backward: you can take what you know about the present and it will tell you what the universe was like in the past.

Schrödinger's
Sabertooth

According to the theory, there's no time limit to this predictive power. It's a bedrock principle that quantum information doesn't disappear; it just gets transformed into new quantum states. That means that if you know the quantum state of the universe today, you can calculate its quantum state at *any point in time*. Quantum mechanics tells us that the universe stretches backward and forward in time forever.

This means something very simple: the universe has always existed and will always exist. If our understanding of quantum mechanics is correct, there is no beginning of the universe.

The Relativistic Universe

Einstein's theory of relativity tells us a very different story. One problem with quantum mechanics is that it usually assumes that space is *static*, like a fixed backdrop where you can hang particles and fields. But relativity tells us that this is very wrong.

According to relativity, space is dynamic in the sense that it can bend and stretch and compress. We can see it bend around heavy objects like black holes or the Sun. And Einstein's theory also describes how all of space is expanding. Space isn't just a flat emptiness; it gets distorted locally by heavy things, and it is growing and getting bigger.

This is a crazy idea that first came from the mathematics of relativity, but we now have experimental proof of it. Using telescopes, we can see that galaxies are being pushed away from us faster and faster every year. Everything in the universe seems to be getting more spread out and colder, like a gas cooling as it expands.

What does this mean for the origin of the universe? Well, if you run the clock backward, our observations predict that the universe was once hotter and denser. And if you look back in time far enough, the universe gets to a special point: the singularity.

Here the density of the universe is so great that the calculations of relativity go a little bonkers. They predict that the universe gets so dense, and that space curves so much, that it reaches a point of infinite density.

This relativistic view tells us that, in a way, the universe did have a beginning, or at least a special moment in time. Everything we see around us, including all of space, came from that point. Unfortunately, relativity can't tell us what happens at that point, but we know that it's *different* from any point afterward. It's like a wall past which relativity cannot explain.

Who's Right?

So our two pillars of modern physics tell us two very different things about the possible origin of the universe. On the one hand, quantum mechanics tells us that the universe is eternal, and that it has always existed. On the other hand, relativity tells us that the universe came from something: a point of infinite density that happened fourteen billion years ago.

We know that quantum mechanics can't be totally correct, because there are things about the universe it doesn't describe. For example, quantum mechanics doesn't describe the force of gravity or the bending of space. But we also know that relativity can't be totally right either, because it breaks down at the singularity and it ignores the quantum nature of the universe.

It's clear that to answer our questions about the origin of the universe, we're going to need a new theory. One that is able to describe the early moments of the universe and unify the best parts of both quantum mechanics and relativity. And maybe once we have this new theory, we will be able to answer the larger questions, like where the universe came from and how it came to be.

What Are the Possible Theories?

While we don't yet have a working theory that unites quantum mechanics and relativity, we do have a lot of different ideas under development, from string theory to loop quantum gravity to crazier ideas with sillier names (geometrodynamics, anyone?).

These ideas generally fall into one of three categories:

(1) Quantum mechanics is mostly right.

(2) Relativity is mostly right.

(3) Neither of them is right.

Let's dig into these possibilities and see what they say about our universe's origin.

Quantum Mechanics Is Mostly Right

One possibility is that quantum mechanics is mostly right, and the universe has always existed and will always exist. Of course, the biggest problem with the quantum picture of the universe is that it doesn't describe how space is growing and changing, and it doesn't describe how the universe came from an extremely hot and dense state fourteen billion years ago.

What if we could keep most of quantum physics, and add a quantum explanation of how space can change? That might give us the answer we're looking for.

Quantum gets spacey

To do this, some physicists have tried to paint a different picture of space. We're used to the idea of space as something fundamental: things exist inside of it, and it allows things to have a location and move around. As far as we know, it doesn't exist inside anything else.

But what if that's not true? What if there's something deeper and more fundamental than space? What if space is actually made up of smaller quantum bits, ones that can sometimes get organized together to have the familiar properties of space?

We see this kind of thing all the time in physics, and we call it "emergent phenomena." For example, liquid water, steam, and ice are all emergent phenomena of the same thing: water molecules and how they interact with one another depending on their temperature and pressure. In the same way, it might be that space itself emerges, and is stitched together from, more basic bits that are the essential units of the universe.

What are these quantum bits of the universe? Each theory is different, but here is what we can say about these bits:

(a) They each represent a location. At each location, you can have particles and fields, and therefore you and other things.

(b) They are not arranged in order. These bits are not sitting neatly in rows somewhere. Instead, they exist as a sort of quantum foam.

(c) They are related to each other by quantum relationships called "entanglement," where the probability of one can affect the probability of the other.

These theories say that what we call our universe is actually a network of these quantum bits connected to each other in a special way.

They also say that what we perceive as "space" is really just the strength of the connections between the bits in the network. For example, quantum bits that are strongly entangled are locations that we perceive to be close to each other. And bits that are weakly entangled are locations what we perceive to be far from each other. In this way, space emerges as the fabric that holds all of these bits together.

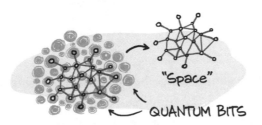

"Space"

QUANTUM BITS

This makes sense from a quantum point of view because it reflects what we see in our universe. Things that are close to each other (heavily entangled) are likely to affect each other, and things that are far away (less entangled) are less likely to affect each other. For example, if a star goes supernova on the other side of the universe, you can shrug it off and enjoy your lunch. But if a nearby star goes supernova, then your lunch is toast (as are you).

Hey, neighbor!

Whoa, where'd you go?

This also makes sense from a relativistic point of view because it allows space to be flexible. It could explain the bending of space as a temporary change in the relationships (or entanglements) between quantum bits that are near a heavy object. And it would explain how our universe can be expanding: new quantum bits are getting entangled with the current network, effectively making more space, which we see as the universe getting bigger.

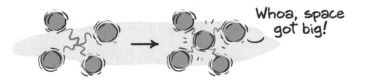

Whoa, space got big!

   This idea may sound crazy, but it gives us a clear answer to the question "Where did the universe come from?" According to this view, the universe came from a larger metauniverse filled with these quantum bits, and what we call "space" is really just a blob of those quantum bits that happen to be connected to one another.

   This idea also has some interesting implications. If our universe exists as a blob of connected bits in a quantum metauniverse, then there could be other universes out there. Our blob universe could exist along with other blob universes, each with different ways in which the quantum bits are connected. It also means there might be a lot of space out there that is *not* connected to any particular universe. There might be quantum bits in this foam that are unconnected, or are connected together in incoherent ways. In other words, there might be a lot of non-universe in existence.

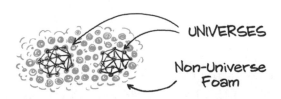

UNIVERSES

Non-Universe Foam

   Of course, even if this idea answers the question of where *our* universe came from, it also raises more questions. Namely, what are these quantum bits? Where did *they* come from? What caused them to form our universe? And where did the larger metauniverse come from?

Relativity Is Mostly Right

Another possibility is that relativity is mostly right, and our universe actually came from a single event in time that happened fourteen billion years ago (i.e., "the singularity"). But how does this match up with quantum mechanics and the idea that the universe always existed?

There's also another problem with relativity and its prediction of the singularity. According to quantum mechanics, a singularity is impossible. A core concept in quantum mechanics called the Heisenberg uncertainty principle says that nothing can be resolved to such a small size. In quantum mechanics, there is a minimum amount of uncertainty that everything must have, and these effects get stronger the more you squeeze matter and energy together. How can this work with the idea of the entire universe stuffed inside of an infinitely small point?

Some physicists have found a few loopholes to these quantum restrictions, and have proposed a few tweaks to the relativistic origin story of the universe.

First, physicists have allowed for the possibility of a fuzzy singularity. Maybe the universe didn't come into being as a single point but as a fuzzy patch of space and time. In other words, maybe the universe has always been quantum, even from the start. This would avoid the pesky mathematical problem of describing a point with infinite density, which relativity also has trouble with.

Second, physicists can get relativity to agree with quantum mechanics' requirement that the universe always existed by tweaking

THE FUZZY NAVEL OF
THE UNIVERSE

what the word "always" means. Relativity's idea of a singularity bothers a lot of people because it represents a boundary, an edge, to time. In a way, it tells you that time ends, and beyond that point there is no more time. But what if time can both always exist and end at the same time?

Stephen Hawking and friends had an idea for how to do this. What if time itself was created in that fuzzy singularity? They called this idea the "no-boundary proposal," and it treats time as more circular than a straight line. In this context, it doesn't make sense to even talk about a time before the fuzzy singularity because time didn't exist. According to the theory, time rotated into being within that fuzzy singularity, and went from imaginary to real. Hawking used a simple analogy to explain this: it's like asking what is north of the North Pole. The fuzzy singularity is like the North Pole of time, and asking what came before it has no meaning.

The start
of Time

All of this tells us that if relativity is correct, the universe didn't come from anything. It means that the universe came from itself, in a way. Both time and space started together, and it makes no sense to think about what came before. According to relativity, the universe is its own origin.

Neither of Them Is Right

The final possibility is that neither quantum mechanics nor relativity is right. Maybe the universe *didn't* always exist (as quantum mechanics requires) and it also never "started" (as relativity suggests).

Sometimes in physics, you get an answer that doesn't really make sense because you've asked the wrong question. For example, asking the question "Where did the universe come from?" assumes that the universe had to come from somewhere. It also assumes that there is an alternate possibility, as if there are conditions under which the universe could have *not* existed.

But what if the universe just is? What if it *had to be*, and the alternative, that the universe not exist, is not really a valid option?

This might sound like kooky philosophical semantics, but there is actually a very mathematical argument that supports it. In fact, it's the most mathematical argument possible: What if the universe itself is mathematical?

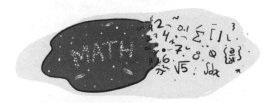

In physics, we use mathematics to describe the laws of the universe. It is the language of physics. But what if math is more than a useful way to count your stars or solve physics problems? What if math doesn't *describe* the universe but *is* itself the universe?

In this view, the universe is a mathematical expression, a raw concept of logic and possibility. It exists in the same way that the number 2 exists, or the equation "3 + 7 = 10" exists. Nobody ever asks, "Why does the number 2 exist?" or "Where does the number 2 come from?" It just . . . is. In the same way, some physicists and philosophers say that the universe exists *because* it works mathematically. All the laws of physics that describe our universe make sense, and therefore they are.

In fact, these physicists imagine that *all* sets of laws of physics that make sense mathematically *must be real and exist*. For example, there might be a set of laws of physics in which gravity is three times stronger, or one that has a fifth fundamental force of nature. If the equations work and the laws don't have any logical inconsistencies, then according to these physicists, that universe must exist. Just like all numbers or all logical equations (like "1 + 1 = 2") exist, so must any self-consistent universe formulation. And if a potential set of laws of physics doesn't work, then a universe with those laws must fizzle out or never come to be.

Can this be real? It might be. Many physicists are skeptical, because currently there seem to be a lot of different ways that one could build a set of mathematical rules for a universe. For example, string theory, a potential theory of quantum gravity, has 10^{500} variations, all of which are consistent with our universe.

But maybe it's just that our theories are unfinished. It may be that once we complete our understanding of the laws of nature, we will find a single valid theory that tells us there is only one possible mathematical universe. In that case, our universe wouldn't just *have* to be; it would also be the only way *to* be.

How Can Something Come from Nothing?

If you came to this chapter hoping we would answer this fundamental question about the universe, you are not alone. Unfortunately, most of these theories seem to tell us that the universe didn't come from anything. They tell us that maybe it always existed, or maybe it *had* to exist, or maybe it doesn't make sense to even ask where it came from.

It could be that this reflects physicists' preference for deflecting the question. After all, if you can show that the universe came from something, you then have to ask: Where did that something come from? The cycle would never end.

Nah, I'd like to retire someday.

But deflecting the question is a little frustrating because it flies in the face of a very ingrained preconception we have about the universe: that everything has to come from something.

From early in our education, and from our everyday experience, we learn that nothing ever comes for free in this universe. We are taught to believe that energy is always conserved and that things don't mysteriously appear out of nothing. There is always a reason, and our human brains have evolved to look for those reasons.

But actually, we've learned in the last few years that even this basic idea isn't necessarily true. When we look out at the universe, we see that space is actively expanding, and new space is being created all the time. That new space isn't empty; it's filled with nonzero vacuum energy. With it, new particles can pop into existence, bringing new energy and matter into our universe.

This means two things. First, the universe is still being born (in other words, it hasn't finished "coming" from anywhere). Second, it is possible for energy to spontaneously appear. It's happening all around us as we speak.

So maybe the question "Where did the universe come from?" isn't the best question we could be asking. The universe exists, and maybe it exists for no other reason than for us to marvel at it and learn from it.

Maybe the real question we should be asking is, "What are we going to do with it?"

Will Time Ever Stop?

E ventually, everything in life seems to come to an end.

Lazy summer afternoons, secret boxes of cookies . . . even punishing winter storms and broken hearts don't last forever. Time ticks forward, and inevitably, both joy and pain fade into the past, making room for the present. The one thing that never seems to end is time itself.

It would be nice to know if time will end someday, or, at the very least, if time can be stopped. It would help you plan out your life, or maybe hit the pause button every once in a while to savor a particular happy or meaningful moment.

But can time be stopped? Will it end one day, or will it keep marching forever toward an infinite future? Will time one day . . . run out of time?

Can Time End?

Unfortunately, there's a lot we don't know about time. In physics, we know that it connects different configurations of the universe. For example, if you throw a ball straight up in the air here on Earth, we know that after some time it will be back where it started. That's what physics is all about: describing how the universe moves forward with the passage of time. The laws of physics tell us what is allowed to happen, and what is forbidden with respect to time.

But can it ever end or stop? The answer might depend on what we mean by time stopping. Let's examine some possibilities.

Does It Mean "No More Laws"?

Time is the thing that orders and connects all the different ways that the universe can be. So when time stops, maybe all the rules go out the window. Since the laws of physics are based on time and dictate what should happen as time ticks forward, perhaps an end of time simply means the death of *order*. Cause and effect might no longer have any meaning, and the universe would exist in a state of complete disarray. The death of order is not something we like to think about.

Does It Mean "No More Changes"?

Or perhaps an end of time simply means that the universe can't change anymore. If time is the thing that allows the universe to change, then when time stops, the universe might . . . *freeze*. Whatever state everything is in (balls flying through the air, lightning striking from a cloud, stars collapsing into black holes, etc.), that state would become fixed. And possibly forever: If time stops, can it just stop for a *little while* and then start up again? That would require some outside clock to be counting the number of frozen moments (more on that later). If time freezes, then maybe *all* clocks freeze, and the universe might never recover.

Does It Mean "The End"?

It's hard to imagine the universe existing without time. Relativity tells us that time is very closely related to space, and that they are better seen as the combined idea of "space-time." Perhaps that means that they are tightly connected, or even part of the same thing. It might even be that the universe existing *at all* is connected to the existence of time itself, and that without time there is no universe. This means that the only way for time to end is for the whole universe to end as well.

All of these possibilities point to a more fundamental question about time and the universe: Can the universe exist without time? In other words, can there *not* be time?

To answer that, let's recap what we know about time.

What We Know About Time

Time as a subject is actually not that well understood in physics. It's so ingrained in our theories about how the universe works that very few scientists have made any progress on the question of whether the universe can exist without it. Think about it: any experiment to test it basically requires time. You have to compare your experiment before and after a certain event, and without time the words "before" and "after" don't make sense. It even takes time to come up with the experiment!

But over time (no joke intended), physicists have gotten some important clues about the nature of time and how it relates to the universe. Specifically, we've learned the following:

(a) Time had a beginning (of sorts).

(b) Time is relative.

(c) You *could* have no time.

Let's dig into each of these clues one at a time.

Time Had a Beginning (of Sorts)

Until recently, most scientists believed that the universe was infinitely old and static. That means that the universe had always existed the way it is and, by extension, would always exist the same way. When we looked out into the night sky, not much seemed to move. The stars changed position a little with the seasons, but they didn't seem to change from year to year or even century to century. That the universe had always been that way, with stars hanging motionless in space, felt like a natural idea.

But when astronomers looked more closely, they made some shocking discoveries. Using techniques that let them measure the distance to faraway stars, they were amazed to discover that some of the smudges they thought were gas clouds were actually entire galaxies. And these galaxies seemed to be almost impossibly far away. Even more amazing, the light from these galaxies was shifting in color, which meant that the galaxies were moving away from us. It seemed that the universe was much larger than they had imagined, and it was hurrying to get bigger.

All of a sudden, we understood that the universe was not a static panorama of stars fixed in space; it was growing and changing. More discoveries revealed that it was getting cooler and less dense.

This gave humans a completely new view of the universe and its history. Because if the universe is expanding and cooling now, then what was it like in the past? If we run time backward, then we can imagine that the younger universe was denser and hotter. But we can't run time backward *forever*.

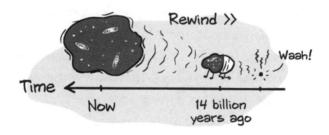

At some point, this backward picture of the universe gets so small and hot that it reaches a wall, a point of infinite density known as a "singularity." This singularity is what we project our universe's past to be, and it breaks down all of our theories about the universe. Even general relativity, which tells us how space curves around matter, can't describe the singularity, where the curvature becomes infinite. We don't know what happens to time and space under such extreme conditions. But it might represent a back end to our timeline of the universe.

In fact, some theories, which try to fuse general relativity with quantum mechanics, suggest that the singularity could be more than just a special moment in time. They suggest that space and time are so intertwined that you can think of this moment as the actual *start to time itself.* In other words, a beginning of time.

And if time had a beginning, could it also have an end?

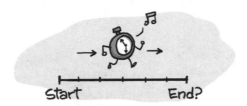

Time Is Relative

We also know that time has a lot of strange properties, none more so than the fact that it doesn't flow at the same rate everywhere. In some places in the universe, time moves faster than in others. It's hard to believe, but physics tells us that there's no central cosmic clock keeping the universe in sync. Instead, every point in space has its own clock, and how quickly or slowly it ticks depends on how fast you are going and how close you are to a heavy mass such as a black hole. If someone is moving by you really fast, you will see their time moving slower than it does for you. And if they are near a black hole, and you are far away from it, you will also see their time moving slower than yours.

There's a common misconception that this means time slows down *for you,* as if you would feel time moving more slowly. But you don't. If you are zooming past someone, or are near a heavy object, other people will see your clock moving slowly, but you always feel time moving normally.

It's all about where you are and how fast you're moving relative to the clock. If you're on board the spaceship with a clock, you're not moving relative to the clock. And if you're near a black hole, then the clock is also there with you. In both cases, the clock will seem to run normally to you. But if you left someone back on Earth, then *they* see your clock run slowly because they are not with you.

Does this mean that time can stop or end? Not necessarily.

At half the speed of light, time on board a spaceship appears to go about 15 percent slower. At 90 percent of the speed of light, time on the ship is just over twice as slow, and at 99.5 percent of the speed of light, it is almost ten times slower than normal. If ten hours have passed on Earth, the spaceship clock would only appear to count a single hour. But while we can make the clock on the ship tick forward as slowly as we want by speeding up the ship, it will never actually stop. To stop time on the ship's clock, it would have to fly *at* the speed of light, which is not possible for anything with mass.*

In the same way, for someone watching you from afar, the clocks on your ship will appear to run slower as you get close to a black

---

* Which, of course, makes you wonder: How does a photon experience time? Flying through the universe at the speed of light, it sees everything else moving relative to it, and so all of the clocks in the universe look like they are frozen in time.

hole. As we talked about in What Happens If I Get Sucked into a Black Hole?, you'll start to look like you're moving in super-slow motion to someone far away. When you reach the edge of the black hole, you'll actually appear to be almost perfectly frozen in time, waiting for the black hole to grow and engulf you. But from your point of view, time will flow normally and the trip into the black hole will be seamless.

So you can't stop or end time by strapping yourself to a rocket and going really fast or going to a black hole. But if you need some extra time to work on your physics homework, just convince your teacher to jump on a spaceship so her clocks run slower than yours. Then you can take your time.

A Black Hole ate your homework?

Yes. You should go check.

You *Could* Have No Time

Time is so basic to our experience that it's hard to conceive of a universe without it. But that doesn't actually mean that it's a core part of the universe. It just means that our thinking may be too narrow or subjective. The history of scientific discovery reminds us to check our preconceptions because our limited experience isn't always universal.

A fish that has lived in a flowing river its entire life can't imagine water without feeling its *flow*, but we know it's possible. The idea of water flow is not a deep and necessary component of the universe, but something that happens under certain circumstances. In other words, there can be water without flow.

What if our entire conception of the Universe is wrong?

Oh, just go with the flow.

Some physicists think that the same thing could be happening with time. It's possible that time is not a basic permanent fixture, but a special condition, just like the flow of the river. To make this theory work, you would need something else, call it "meta-time," from which regular time emerges. This meta-time can flow like time, or it can . . . just not. When this meta-time flows, we feel the effects of time. And when it doesn't flow, we feel that time ends.

It might be that some basic rules that we think are absolute and necessary, like causality and time only moving forward, are only special cases of this meta-time flow. Maybe this meta-time can do other things, like form the equivalent of whirlpools or waterfalls, and we would see time moving in loops. Or maybe it can break causality and you can eat dessert before dinner.

Of course, that doesn't mean there are *no* rules or that anything goes. This meta-time still has to have some similarity to our idea of time or it wouldn't be possible for time to flow at all. It still has to follow some rules, and if it does, then those rules might dictate a situation where time as we experience it can stop.

What this means is that time (as we know it) doesn't *have* to exist, and that there can be a universe without our familiar kind of time.

We don't have any evidence to suggest this is actually our reality.

But it's not *entirely* speculation. We know that our understanding of space and time breaks down fourteen billion years ago when the universe was very hot and dense, which leaves an opening for us to consider creative ideas.

How Would Time End?

At this point, we are well beyond the comfort zone of physics and into a region where we have to start guessing. But that's just how science works. New ideas about how the universe works aren't usually born as complete mathematical concepts all at once. Instead, they develop step-by-step, with the pieces coming together gradually over years, decades, or centuries. We sometimes explore wild paths until a coherent picture that can actually be tested by experiments takes shape. It's like building a house of cards, not from the bottom up but by holding each card in midair until you assemble the other cards around it.

What we know so far suggests there are several ways in which time can end.

The Big Crunch

One possible way for time to end is to mirror the conditions in which it began. We think that time might have started when the universe was hot and dense and space was unimaginably compressed in the Big Bang. What if the universe somehow returns to that state in a *reverse* Big Bang? Would time end then?

Big Bang   Big Awkward Teenage Years   Big Crunch

In fact, it might. We know that the universe expanded rapidly in its first moments, and has continued to get bigger in the billions of years since. That expansion has accelerated, so that galaxies are moving away from us faster every year. But we don't understand what's causing that acceleration. We call it "dark energy," but a cool-sounding name doesn't really tell us what's going on. And since we have no idea what is expanding the universe, we have almost no way to predict what it will do in the future. For example, that acceleration could stop and then reverse. Instead of increasing the speed at which other galaxies are flying away from us, it could slow them down, and eventually stop them and turn them around. Instead of stretching space into a bigger and bigger universe, this force may compress it, hurtling these spinning galaxies toward a massive cosmic collision known as the "Big Crunch."

What would happen if all of the matter and energy in the universe is compressed again into a tiny space? The truth is that nobody knows. It would be just like the conditions of the Big Bang, which are also a mystery to us. But that doesn't stop us from having fun thinking about it!

It could be that time will simply end along with the rest of the universe. It doesn't have to be an abrupt end. It could be a curved end in the same way that, for example, the direction of north ends at the North Pole. Time would just be capped at that point, and there would be no more time beyond that.

It could also be that space and time continue, even if all the matter and energy of the universe gets crunched into a singularity. Causality and the rules of our universe would continue to operate, but everything would be weird and unfamiliar, without the particles or the forces we are used to. In that case, time wouldn't end, even if the universe is unrecognizable.

Or it could be that the singularity creates another Big Bang and a whole different universe comes into being. This new universe could still be connected to our universe through a thread of time, which means that time doesn't end; it just starts again. If this is true, then that thread of time would connect an infinite number of universes both forward and backward in time.

The Heat Death

Another way in which time could end is through sheer boredom. To understand this, we need to think about why time marches forward in the first place. It seems like something is turning the crank on the universe's internal clocks, and is only turning it in one direction.

This is something physicists have puzzled over for a long time, even before there were physicists.* It strikes them as odd that time has two directions, but it only goes one way. There must be something, physicists argue, that makes time go forward instead of backward, like a deeper engine to which time is shackled.

THE GREAT HAMSTER OF TIME

And some physicists think they have found it. The universe does have a sort of built-in direction marker: entropy.

It's easy to misunderstand entropy, and to confuse it with general messiness or disorder. But that's not necessarily the case. We say that something has more entropy if there are more ways to arrange its internal particles. For example, if you require that a bunch of matter be clumped together in a corner, there are fewer ways to arrange it than if you let the particles spread out wherever they want. The same is true for temperature: there are fewer ways to arrange a blob of stuff if it has to have hot spots and cold spots than if it's all even temperatures, where any particle can be in any spot.

---

* Physicists don't like to think about this, the BP Era.

One funny thing about entropy is that it steadily marches upward as time goes forward. Our universe started with very low entropy, squeezed as it was into a very organized, dense state, and it's been expanding and gaining entropy ever since.

But the other fascinating thing about entropy is that it also has an edge: a state of *maximum possible entropy*. When everything has cooled and spread out completely evenly, entropy has hit the ceiling and can't go any higher. More importantly, it can't go any lower either. Once all the sand in the hourglass has fallen to the bottom, it can't flow back up again and the universe is stuck.

What does that mean for time? This state, known cozily as the "heat death of the universe," means that it's no longer possible for the universe to do anything useful. Most of the things you want to do (make a planet, charge your phone, run a lap) require energy to flow, which is only possible if there are places where it's imbalanced or concentrated (like your phone battery). But if all those imbalances have been smoothed out and everything has reached maximal entropy, then you can't do anything useful. Energy can't flow, like water in a perfectly level and still puddle. You made it to the end of the universe, but there's nowhere (or nothing) to charge your phone.

Some physicists look at the correlation between time and entropy and are tempted to say that time flows forward *because* entropy increases. The second law of thermodynamics says that entropy and time always increase together. These physicists suggest that if entropy reaches a maximum, then time itself will also stop!

Of course, this seems like a big leap, since (a) we don't know if entropy actually drives time forward, and (b) maximum entropy doesn't mean that the universe stops moving. Even at maximum entropy, particles can still fly around. The only limit is that they can't *increase* (or decrease) the overall entropy. It might be that the universe continues at this state of maximum entropy, but time keeps flowing.

It would certainly *feel* like the end of time, though. At maximum entropy, the universe will sit as a bland puddle, and nothing interesting can or will ever happen again. So while it may not be the end of time, it will definitely be the end of fun.

PARENTING AT THE END OF TIME

Who Knows?

If time is not a fundamental property of the universe, but just something that happens under some special conditions of a "meta-time" (like the flow of a river), then it's possible for those conditions to end.

Maybe we'll reach the end of our river in meta-time, causing time (as we know it) to unravel so it no longer ticks forward. The universe could then exist in a state without time, like a river without flow (or a lake). This new state would be wildly different than anything we have ever experienced or imagined. Without time and space, events in physics would not be causally connected. The universe would just exist as a frothing mass of unconnected quantum randomness.

To understand this requires us to know how quantum mechanics and space play together, a theory that physicists since Einstein have been fruitlessly searching for. This means we can't even fathom how it would work, or what might cause all those conditions to change. For all we know, it could happen tomorrow or the day after. Only someone with an outside view of the flow of this meta-time would know.

But this end of time could also be temporary. Like a lake that then feeds a different river, the meta-time could still evolve and pull the complex threads together to make time flow again.

Interestingly, we might not even notice if time stops and starts again. We measure time by physical processes that step forward regularly: a ticking clock, sand falling down an hourglass, electron atoms jumping between states, and so on. So if time unravels or stops, then those clocks would also stop. That also includes you, since you are a physical being. And since your thoughts and experiences can only happen if time moves forward, you wouldn't be able to tell when the flow of time stops or slows down. Like a character in a paused movie, you would have no idea how many times, or for how long, you were frozen.

The End of Our Time

It's time to admit the truth: we don't really understand time. Somehow, like the mysteries of our own minds, living in time doesn't necessarily give us insight into how it works.

We do have some beginnings of ideas. It might be that time is eternal, and the universe's clocks will tick forward forever into the infinite future. It might also be that time is not fundamental to the fabric of the universe but a special arrangement that might not last forever. Or it might be that time *is* fundamental to the universe, and the only way for time to end is for the universe to cease to exist.

Right now, time seems to be flowing smoothly. But who knows, maybe special circumstances like the Big Crunch or the heat death will reveal something new.

We might be wondering about this until the end of time.

Is an Afterlife Possible?

S adly, everybody dies.

We're all stuck with that terminal condition called human life, which means our bodies won't last forever. Eventually, they will stop working, and our physical selves will give way to entropy and decay. But is the end of biological life also the end of *you*?

I guess a meteor's a pretty cool way to go.

This might be the deepest and most ancient of questions: What happens after we die? It's a question with so much emotional resonance that it's at the heart of most religions and cultures. The variety of ideas about the afterlife is truly impressive, and sometimes even a bit nutty. For example, we all go to a giant banquet hall in a huge space tree? C'mon, Norse mythology.

Usually, this is a topic that scientists leave to philosophers and religious scholars. But we have learned a lot about how the universe works since we started thinking about this question thousands of years ago. Is an afterlife possible given what we know about the laws of this universe?

The Physics of Heaven

There are lots of ideas about what life after death might be like. In most religions, this involves you existing in some new, non-earthly situation. What this new situation is like depends on the religion: sometimes it means existing among the clouds with angel wings and harp recitals (or the opposite: in a dark underworld with pitchforks and fire), and sometimes it means riding around with the Sun or singing endless songs over mugs of ale with warrior-gods. In most cases, this afterlife lasts forever, making everyone a little jittery about what kind of accommodations you'll have once you get there.

Also, in most cases, you're somehow still *you* in the afterlife. Your individuality, consciousness, and memories all somehow survive, making it possible for you to experience and be self-aware in this new phase of your eternal existence.

Is any of that possible, scientifically? Could you somehow be transmitted to another realm where you could continue to exist, as you, but now with a toga or extra-comfy slippers? Let's take it at face value and think about how that might work. Scientifically, these are the three key elements that seem to define a traditional afterlife:

(1) There's a *you* that can outlive your physical body.

(2) That *you* is captured and transported to another location.

(3) You exist in that other location, still able to experience things, forever.

Let's think about each of these elements one at a time, and see whether there's a version of these ideas that's compatible with what we know about the physical universe.

A *You* Beyond You

Most religions assume that there's a part of you that can survive the death of your body. The first step to understanding whether any of this makes sense scientifically is to figure out what part of you we are actually trying to preserve. For example, most of us are not interested in continuing to inhabit our own deceased bodies, staggering around like zombies and grossing out all of our former friends.

So if we are willing to let go of our physical bodies, then what is it that we want to preserve? What, exactly, makes you *you*?

This is the kind of question that science can really dig its teeth into. Physics operates in the realm of the physical (duh), and so it

assumes that everything follows the laws of physics. And as far as we can tell, what makes you *you* is just . . . your particles. More specifically, the *arrangement* of your particles.

You see, it turns out that everything we see in our world is made out of the same building blocks. All the matter that we interact with is made out of two types of quarks (an "up" and a "down" quark) and electrons. And that's it. The two types of quarks can combine in different ways to make neutrons (one up quark + two down quarks) and protons (two up quarks + a down quark), which then combine with electrons in different proportions to make every element in the periodic table. And from those elements you get everything from llamas and boats to microbes.

In other words, there's no difference between you and every other thing (or person) in this world except for how those elements and particles are arranged together. A kilogram of you has pretty much the same particle content as a kilogram of lava, or ice cream, or elephant. If you were to write a cookbook for making anything on this planet, every recipe would have the same list of ingredients: a three-to-one ratio of quarks to electrons.

A PHYSICS COOKBOOK

But as anyone who has failed in the kitchen knows, a recipe is more than a list of ingredients. If you mix the raw materials the wrong way, you can get something that not even your dog will want to eat. In the case of you, the crucial thing that differentiates you from lava and ice cream and bugs is the *arrangement* of your particles, not the particles themselves.

In fact, there isn't even anything special about the exact particles that you're made out of. From the point of view of physics, all electrons are the same. If you swapped out your quarks or electrons for a fresh set and put them back exactly where your old ones were, nothing would change.

This means that you are just the *information* about how those particles are arranged, which means that you *can* survive even if your body dies. All you have to do is somehow copy that information and have it live on somewhere else.

Transporting You to Another Location

The next step in most afterlife scenarios is that you (whatever it is that makes you *you*) are somehow transported to another realm or location. From a physics point of view, this means that your information is somehow copied or transferred away from your body and taken to another place. But this raises several important questions:

✦ How is this information read or captured (or raptured)?

✦ Does all of your information get copied, or just part of it?

✦ Which version of you gets to go on?

The first question, "How is this information read or captured?" is really more of a process question. Whatever mechanism takes you to the afterlife must be based on some physical principle if it is to work within a logical universe. To date, we do have technology that can scan your body, like an MRI or a CT scan. And we have technology that can detect individual atoms. Both of these technologies are getting better every day, which means it's not inconceivable to think that someday soon there will be a process that can scan your body down to the atoms or particles.

But from a physics point of view, there are two problems you run into. First, any scanning would require imparting energy on your body. To detect individual particles, you would need to see them somehow, and typically that means hitting it with a photon or some other particle. On top of that, the universe doesn't allow you to copy quantum information without a cost. This is a core principle of quantum mechanics called the "no-cloning theorem," which says that quantum information can't be copied without destroying the original in the process. So far, we haven't seen evidence that people are getting their body scanned or that their particles are getting destroyed at the quantum level when they die.

We're also not sure you *can* copy all your particles at the quantum level. Scanning all of the quantum states of a human body is no small feat, since there are 10^{28} particles, which dwarfs the total computer memory of human civilization, currently around 10^{21} bytes. Today, we could maybe store the information contained in one of your toenails if we used every computer in existence.

Of course, there is the possibility that this information could simply be available to whatever is executing your passage to the afterlife. Perhaps our universe is like a simulation running in another universe, in which case your information simply lives on a hard drive somewhere, ready to be read and copied.

The second question, "Does all your information get copied, or just part of it?" is a more philosophical question. For example, does the afterlife really need *all* the information in your body? Is it really important, for example, to know what every quark in your toenail is doing at the moment of your death?

Or could it be that the afterlife only needs part of your information? If so, which parts are needed?

We know that the arrangement of all your particles is what makes you unique, but it might be useful to think about what that arrangement does. Your arrangement of particles defines a biological machine, a set of mechanical processes at the cellular level that takes in information from the world and reacts to it with certain actions. Is the arrangement of quantum particles in your toes or even your limbs necessary for that? What about your gut? Do you still need your gut instincts in the afterlife?

It could be that what you actually need for the afterlife is not the arrangement of every particle in your body but just the *design* of your biological machine. Maybe what lives on is not the quantum information of all your cells but how those cells are connected to one another, and what information they have stored in your brain circuits. That would definitely help you save on hard drive space.

And you can imagine compressing your *you* information even further, ignoring more and more details like a fuzzy JPEG image of yourself. Would that *you* still be you, though? Or would it just be a simplification, like the "essence" of you?

The last question, "Which version of you gets to go on?" is more of a timing question. Our bodies and our minds change a lot over our lives. While our conscious experience and knowledge get bigger as we age, our bodies and mental capacity peak and start to decrease at some point. Which version of you is the one that goes to the afterlife? In other words, when does the copying and pasting take place?

If it happens at the moment of your death, you might find something to complain about. What if you are not at your best when the moment happens? Or what if there are things that happen to you leading up to your death that you'd rather not carry with you to eternity? Who gets to choose, and how?

Or perhaps the process of capturing you for the afterlife happens on a curve. Maybe what gets copied is an average of you, or a sum total of what makes the JPEG image of you unique. If all we are is information, science knows a lot of tricks you can use to compress, average, or find the most important features of that information.

Existing in Another Location Forever

The last piece of the afterlife puzzle is the idea that your you-ness somehow lives on forever in another place. In some ideas of the afterlife, this place is among the clouds (or down below the earth). In other cases, it just exists as a separate realm, detached from our plane of existence.

This idea sounds fantastical, but the notion of multiple universes is something that physics actively considers. Whether that's at all plausible depends a lot on where this afterlife is located.

It could be that our universe is actually a subset of a larger "metauniverse," an idea physicists conceived to try to explain the origin of our universe. Physics has made some progress in understanding the rules of our universe, but we haven't made much progress in figuring out *why* our universe exists. One concept is that maybe our universe is just a bubble in a deeper, larger universe (the metauniverse), and that our space-time is just a fluke arising from special conditions but not fundamental in itself. In this case, going to the afterlife means that somehow our information gets scanned and copied to that outside universe.

Another possibility is that the afterlife is in a parallel universe. The idea of the "multiverse" in physics says that maybe our universe is not the only one, and that there could be other pockets of space-time elsewhere. In some theories, these other universes are alternate versions of ours, perhaps split through quantum decisions or through different initial conditions or even different laws of physics. If that's the case, there *could* be a version of our universe that is more ideal or utopian (a Shangri-la of sorts). At the same time, there could also be worse versions of our universe, full of fire and fury (a Hades of sorts). Somehow, our information would have to find a way to make it to these other universes, which is something that physicists currently don't think is possible.

Whatever the case, it's interesting to think that the rules of those other universes could be totally different from ours. What sort of adaptation would you have to make to your *you* information to exist in this universe? Would time and causality even work the same way? What sort of vessel or machine (biological or not) would your information sit on? After all, if the afterlife is forever, you also want to be able to think and change and experience things in your new universe home. You want *life* after death, not sitting around frozen forever. That means that the metauniverse would have to be able to run your software, whether through states of quantum objects or something else we can't imagine yet. Think of it as porting a human program over to a new kind of alien computer.

File extension not supported

Uh-oh.

Heaven on Earth

A final possibility is that maybe our universe *is* the metauniverse. It could be that the afterlife can exist within our universe, not outside or next to it.

For example, maybe some neighboring alien species has created a Valhalla for us, and is standing by with scanners ready to copy us into it when we die. Or even more interesting, it could be that *we* build an afterlife for ourselves.

How would this work? Well, we could develop the same kind of technology that we imagine some heavenly force might use to whisk us off.

For example, we could develop the technology to scan a full body down to the molecular or particle level (or at least the human-essence level), and we could also develop the bioengineering or 3D printing technology to build new bodies for ourselves. Those two technologies could be used to create new copies of ourselves that are younger or healthier, and they could be sent to exist in a different location. Maybe we could house them in a more idealized place, or a worse one, depending on why you made the afterlife.

Of course, we are very far from having the technology to do this, and there are also a few tricky quantum mechanical issues to consider, as we mentioned before. In that case, it might be easier to not have a physical body at all!

Instead of trying to re-create your body, you could take advantage of the fact that you are just information and live on in a *simulated* afterlife.

All of the essential information that makes you *you* could be uploaded into a computer, which would then run a simulation for your digital self. Your digital copy would exist in this environment and even grow and change within it. And since it's all made up, this afterlife could be specially tailored for you. Do you want fifty sundaes for breakfast every day? No problem! Do you want to live out your 1980s fantasy or hang out with Jon Hamm *Black Mirror*-style? In a digital world, anything is possible.

Would it last forever? Well, it would last as long as someone keeps the computers plugged in. Interestingly, you could set the time rate inside the simulation to whatever speed you like. Depending on how fast the computer processors are, you could spend a million lifetimes inside your digital afterlife in the time it takes the computer technician to go get a cup of coffee.

Our current computers are not powerful enough right now to store all of your information or mimic the world perfectly, but they are improving quickly, and it seems likely they'll be able to pull off a pretty sweet afterlife in the near future.

Loading rainbows...

Ripples Through Time

As you can see, heaven is no joke. To put on an afterlife requires nothing short of building entire universes, figuring out remote and spontaneous scanning of countless quantum particles, and finding a way to move all this information around without anyone ever noticing. While we can't technically rule it out, from a physics point of view it seems like a tall order.

In the end, all that physics can do is look at the world around us and draw conclusions from what we can test and observe. So far, our current view of the universe is that it follows strict rules. There seem to be no exceptions, no matter how much our minds would like it to be otherwise. To the best of our knowledge, there is no evidence that anything but entropy happens once we die.

Does that mean that physics gives a thumbs-down to an afterlife existing in our universe? That when we die, who we are is gone forever?

Not entirely.

According to quantum mechanics, quantum information can't be destroyed in this universe. What this means is that when your body dies, the particles it's made out of may separate and scatter, but their quantum information won't disappear. That quantum information will probably be absorbed or transformed into other particles, but it will never go away. It will stay encoded in the quantum state of the universe, like an imprint or a clue. Technically, someone in the distant future could examine that imprint and reconstruct who you are and what you did. Such is the power of quantum mechanics.

And this idea extends to your actions as well. Every action you take causes interactions with other particles, and changes their quantum state in a unique way that in principle stores the information of that interaction. In a very real sense, our actions ripple through time, never lost and always present in the quantum history of the universe.

In this way, everyone who ever lived is still with us, through the faint but indelible mark we all leave on the things around us. One day, you too may die, and you will become part of the universe's record. There's an old adage that says we live on in the hearts and minds of those who knew us. According to quantum mechanics, this isn't just true; it's a mathematical fact.

THE UNIVERSE REMEMBERS

Do We Live in a Computer Simulation?

Is this really happening? For real?

That's a question that people often ask themselves whenever they're experiencing something great (or not so great), or sometimes even when they read the news these days. The world we live in can seem so outrageous or so mind-boggling that it's hard to believe it really exists.

Then again, maybe it doesn't!

An idea that has been around for thousands of years is that the universe we live in, the one we experience with all of our senses, might not actually be real. Ancient religions often talk about our world being nothing but an illusion, and Socrates wondered if we would even be able to tell the difference. More recently, Keanu Reeves summed it all up in *The Matrix* with a single word: "Whoa."

We grow up assuming that what we see and feel is what's really out there, and that the universe is full of physical things that are moving around and banging into one another to make the sights and sounds that we take in with our senses. It certainly *feels* real. But *feeling* real and *being* real are not necessarily the same thing. For example, dreams can feel real when they are happening, but that doesn't mean you were actually being chased down the street by a building-size cookie.

Surprisingly, the question of whether our universe is real is something that modern physics has started to wonder about. Could our world not actually be happening? Is it possible that what we experience is just some elaborate simulation of a universe cooked up in an impossibly vast and powerful computer? And most important, how would we know?

Why Even Consider This?

The idea that the world isn't real and that we are actually living inside of a simulation might sound crazy to you. How could our messy and ultra-detailed world possibly be generated by a computer? Even something as simple as a fly buzzing around your living room is dense with detail, from its tiny wings furiously beating on billions of air molecules to its glittering eyes reflecting your face in each of its facets. Could a computer simulate all of that?

Actually, yes. Computer graphics have become incredibly realistic. Compare the simplicity of the original *Toy Story* movie with the most recent sequel (*Toy Story 4?*), and you'll start to understand the huge leaps in computer technology that have been made in just a few years. Virtual reality and video games have also become shockingly sophisticated compared to the blocky polygons of early versions. The latest sports video games are so convincing that it's hard to tell without taking a closer look whether it's a simulated game or actual footage of a live event. The celebrations, frustrations, and tantrums are all there! Considering the rate of progress, it's not hard to imagine that one day it might be hard or even impossible to tell the difference between virtual reality and actual reality.

So real!

Famously, some people even argue that it's *likely* that we are living in a simulation. As we see our technology improve, we start to imagine a future in which everyone is running a simulation of a universe in their home computers. And some even imagine that in those simulations there could be simulated people inside of them running *more* simulations (a simulation inside a simulation!). If you keep going, pretty soon there are way more simulations running around than there is an actual universe, and it makes you wonder: What are the chances that we are living in the *one* true universe, and not in one of the countless simulated ones? Statistically speaking, you would have to put your money on the idea that we are living in a video game.

Philosophically, there is another reason to suspect we might be living inside of a simulation: our universe seems to work *like* a simulation.

You see, our universe has a lot in common with the computer programs that we use to build virtual games and virtual worlds: it seems to follow rules.

The whole project of physics is to uncover the rules of the universe. And the universe does indeed seem to follow them. From quantum mechanics to general relativity, we seem to be inching closer to discovering the source code of the universe. But an often overlooked question is: Why does the universe follow rules anyway? Why is it so consistent and regular?

The laws of physics seem to work everywhere, all the time, in exactly the same way. It reminds you of . . . a computer program. Just like a piece of software, the universe we live in seems to be chugging along, blindly applying a set of instructions set down by a master programmer.

The fact that our universe has an awful lot in common with how

you would expect a simulated universe to run is a fairly strong argument that it could be just that.

Is It Possible, Though?

What would it take to actually simulate an entire universe?

It's clear that programmers are accomplishing amazing things recently, but that doesn't mean it's now easy to build a virtual universe. It's a big leap from describing a simple fly in a single location to describing *everything*. It feels like an overwhelmingly impossible task because "everything" is a lot of things. Not only are there a lot of details in flies and blades of grass, but there are a whole lot of flies and zillions of blades of grass. And that's just on our planet!

To get a sense of what it would take, let's paint a picture of how a simulated universe might work. In our view, there are three basic ways in which it could happen.

Brain in a Vat

In one scenario, a computer is running the simulation and feeding messages to a real human brain. That brain is building its concept of the world from what it perceives through its senses. But those signals are not created by any sense organs in a real body but by the computer simulation. Inside the computer is a model of a whole fake universe that interacts with the brain. When the brain sends a message like "walk forward," the computer simulates the act of moving forward and computes how the world would change and what new inputs to give to the brain.

Alien Brain in a Vat

In a slightly wackier scenario, a computer could be running the simulation for an alien brain, and then pretending that the brain is actually human. The alien in the simulation might think that their brain is a blob of jelly filled with billions of neurons firing at each other, but really, it could be anything. Their actual brain could be much bigger, or smaller, or work on completely different principles, like a vast network of hydraulic pumps or tiny quantum computers or something even crazier.

You Are a Software Program

Prepare yourself for the deepest level of meta. What if we don't have a real brain at all? What if all the brains in the simulation are *also simulated*? In this scenario, all living and conscious minds are part of the larger program. The last few decades have seen huge leaps forward in artificial intelligence, and we are now capable of making computer systems that can mimic the brain's functions of learning and memory and solving problems. As these artificial minds get more complex, they accomplish things that humans confidently assumed AI could never do: defeat the world human chess champion, steer a car through traffic, identify faces, maintain a realistic conversation. It's not hard to imagine creating a virtual world with virtual intelligent beings running around inside of it.

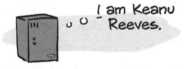

Of course, no matter what flavor of simulated universe you create, you'll still need a massive computer to make it work. To simulate a universe, you have to start with an initial setup: where all the objects are, and how fast they are moving. Then we apply the laws of the universe: What happens to those objects in the first moment? Do they bounce off one another, or pass through one another, or speed up or slow down or turn left? Each object is updated according to the rules, and time ticks one step forward. Then you repeat and see what happens.

This can take a lot of computing power if you have a lot of objects. For example, each object requires some computer memory to keep track of where it is and what it's doing. Now imagine how much memory you'd need for an entire universe, and how much processing power you'd need to crunch through all that data. You would have to simulate every particle and planet in the universe at the same level of incredible detail. Wouldn't that be impossible?

Maybe not. To be convincing, a simulated universe only has to seem real to the beings experiencing the simulation. Here are some ways in which you can get away with less computing power than you might think.

Shortcut #1

The first shortcut you can take is to make your simulated universe a simpler version of the actual universe. For example, you could build it with fewer dimensions than the real universe, or with simpler rules, or with more pixelation. Just because a simulated universe is simpler doesn't mean that it wouldn't seem real to the simulated creatures that live inside of it. It might be that our universe is very simple compared to the real universe, but we don't know any different, so we are happy with the realism that it provides. We could, for example, be like sentient characters in a Super Mario game, and think that this universe is as complex as it gets.

I think, therefore it's a-me, Mario!

Shortcut #2

You can also save on computer power by not making your simulation real-time. There are no rules that say that a simulation has to run at the same rate of time as the real rate outside the simulation.

For example, you could make your simulation run slower, so that one year inside the simulation actually takes one thousand years in the real universe. Then your computer would have enough time to render as much detail as you need to convince the beings inside of it that it's real. They wouldn't know the difference because that's the only time rate that they know. You could even pause the simulation, forget about it, and not restart it until the next day, and whatever is inside the simulation wouldn't notice. For example, do the characters in a video game notice when you've paused it to visit the bathroom? No, because they are *in* the game.

Shortcut #3

A third way to make a universe simulation possible is to be clever about the way you program it. Do you really have to simulate *all* the individual particles in the universe to fool its inhabitants into thinking that it's real? One common trick we use when writing simulations is to zoom in only when you need it. For example, when engineers simulate traffic patterns, they use cars as the building blocks and not the particles inside of each car. And when meteorologists write hurricane simulations, they start from clouds or water droplets, not protons.

In the same way, you could program a simulation of the universe in large chunks, like a rough version, and only go into the particle-level details when it's needed. Faraway planets would only be simulated if someone in the simulation builds a powerful enough telescope to see them, and individual particles would only be simulated when pesky simulated particle physicists build colliders to study them.

Could You Tell?

All of this is to say that it is entirely possible that we (or at least you)* could be living inside of a simulation. Technological trends point to it as a possibility, and philosophy tells us that a simulated universe would be just as valid to us as a real one. Does that mean that we are trapped in this limbo of not knowing? Is there any way to tell the difference between a real universe and a fake one?

It depends on how well the computer has been programmed. If it runs perfectly, then by definition it might be impossible to distinguish it from reality. It could be that the real universe outside of this one is vastly more complicated, and that building a computer powerful enough to simulate every detail that we experience is possible. In that case, we might never tell the difference.

But if computer programming in the real universe is *anything* like programming in our universe, then there is *always* a bug some-

---

* After all, it's possible that we are not real.

where. And that's our best chance to figure out whether our universe is a simulation: to find a glitch.

What would a glitch look like? It would depend on how the simulation is programmed, which makes it very difficult to predict. But we can make a few guesses!

It might be that the simulation has limits in its computing power. For example, it might have a hard time simulating things that happen across vast distances of space. When we build simulations of large, complex objects, we tend to simplify them by dividing them into smaller pieces. It's more feasible to simulate each piece separately and then stitch the results back together. A fake version of our universe might simulate each galaxy as a separate object, so that what happens inside one galaxy is independent of what's happening inside another galaxy. This is like taking a shortcut and hoping that it doesn't make a difference, because the beings in both galaxies are not likely to interact with each other.

But that only works if what happens in Andromeda stays in Andromeda. If there's something in the Andromeda Galaxy that can actually affect what happens in our galaxy, we could use that to try to find a glitch. For example, what if the supermassive black hole at the center of Andromeda is shooting particles at us that we can detect in our atmosphere? That would directly link the two galaxies, and the simulation might not get this right. There might, for example, be irregularities in the trajectory that the particles took to get here, or they might have inconsistent energies. Something like that might tell us there's something amiss in this universe.*

Another possibility is that the universe simulation could have limits in its resolution. Just like the old x86 computers were only able to render blocky, pixelated images in those black-and-green video monitors, it's possible that there is a minimum resolution that a fake universe can simulate. If we were to drill down into space and matter, and find that the universe is pixelated at a level that isn't explained by the laws of physics, that could be a sign that we are in a simulation.

One last possibility is that the simulation we're in could just not be well constructed. It happens all the time in programming here in our universe. As well-meaning or as careful as programmers can be, it seems that the simulations we make always break down at some point. Maybe there are cases that the programmers of our universe didn't account for, or loopholes that they didn't predict. The same thing could happen as we learn more and more about our

---

* In fact, physicists *do* see high-energy particles hitting our atmosphere that can't yet be explained by any astrophysical source.

universe. For example, we have two competing theories about the nature of reality (quantum mechanics and general relativity). Those two theories don't often interact, so they still seem to work on their own, but there are situations in which they totally contradict each other. One of these situations is inside of a black hole, where one theory predicts a singularity while another predicts a blob of uncertainty. It could be that whoever made our universe simulation didn't think the rules all the way through, and were either sloppy or lazy (or rushed) when they built it. Finding out that there is an inconsistency might tell us that there's something not quite right about this reality.

Why Build It?

The biggest question about this whole crazy notion of a simulated universe is, of course, "Why?"

Why would anyone (or anything?) go to all the trouble of creating a whole fake universe and populating it with either connected brains or artificial, sentient beings? Could they be mining us for energy, or enslaving us for some strange purpose?

It could be that our universe is an experiment of some sort. Maybe someone constructed our universe to try to answer a scientific question (such as "In how many universes do bananas evolve?"), or maybe a psychological one ("In how many of those universes are people smart enough to eat them?"). Or maybe we are an experiment for a certain type of universe, and there are countless other universe simulations in which the laws of physics are different, or even the nature of reality is different (that Super Mario world could be totally real in the next universe over).

Alternatively, maybe they're just doing it for fun. What if we're just the equivalent of a fishbowl in their universe, or a toy for their kids? Or worse, what if we're the screen saver in their mega-complex laptops? Who knows what anyone or anything smart enough to build a simulation as complex as our universe will find entertaining?

To summarize, it could be the case that we are all living inside of a simulated universe that runs like a giant machine, governed by rules that we are bound to follow and that we don't yet fully understand, with the possibility that we may never know the true nature of this reality. If this sounds a little grim, then think about the following question for a second: Is that any different than if we were in the real universe?

Maybe the real illusion is that there is a difference between a simulated universe and the real one. From a practical point of view, would it really affect your experience or your sense of self? Perhaps we should be happy to exist at all, simulated or not, and content with the quest to learn all the rules of our existence, whether or not we ever find the answers. If this *is* happening (even in a simulation), doesn't that make it real?

Why Does $E = mc^2$?

f there's one physics equation most people know, it's probably $E = mc^2$.

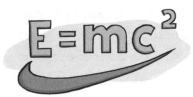

It's the most famous equation in physics, likely because it's easy to remember. Its form is simple and elegant, almost like the Nike "swoosh" logo. Compared to other physics formulas that look more like Egyptian hieroglyphics,* this one definitely has brand appeal. Of course, it doesn't hurt that it came from Einstein, whose

---

* One version of the Schrödinger equation, for example, looks like this:

$$i\hbar \frac{\partial}{\partial t} \Psi(\mathbf{r},t) = \left[\frac{-\hbar^2}{2\mu} \nabla^2 + V(\mathbf{r},t) \right] \Psi(\mathbf{r},t)$$

brilliance (and famous hairdo) have been a part of popular culture since the last century.

But physics formulas are not just math; they're supposed to describe something about the physical universe. And this is another reason why $E = mc^2$ sticks in people's minds. Here E stands for energy, m means mass, and c is the speed of light in a vacuum, or 299,792,458 meters per second. To have them all in a simple, easy-to-remember formula implies that they're connected to one another in a deep and profound way.

But what exactly does that mean? How are mass and energy and light actually related to one another? And what does this relationship say about the fundamental nature of ourselves and the universe?

Mass and Energy

For most of us, mass is the stuff we're made out of.

If something has mass, it generally means that it's heavy, hefty, substantial. We tend to think of things with less mass as lighter, ethereal, or barely there.

This is something we develop in our intuition at an early age, and it's something that was captured by Newton's laws of motion. For centuries, $F = ma$ held the top spot as the most important physics equation in the world. In this formula, F is the force that you apply to an object, m is the object's mass, and a is the acceleration, or how quickly the object starts to move. If the object has a lot of mass, then it takes a really big F to get the object moving. And if m is small, then a gentle push is enough to make it go.

To us, mass is a measure of the *substance* of something. Things with more mass, like mountains and planets, feel more real and solid.

On the other hand, we tend to think of energy as something completely different. We associate energy with heat, light, fire, or motion. It seems like something ephemeral that can flow or be transmitted. It gives you the power to do things and burn things. Like a magical quantity, it's something you can store and release when needed.

For a long time, this intuition about mass and energy fit quite neatly with Newton's laws and our basic understanding of the universe. Mass and energy were two different things, although it was clear they could interact with each other.

For example, if you added energy to something, like a cup of water, you could think about it speeding up the little water molecules in the cup, but not changing the mass of the water. After all, adding heat didn't change how many H_2O molecules there were; it just made them wiggle faster. At least, that's what we thought.

In the late 1880s, physicists started asking pesky questions, like "Where does mass actually come from?" and "What *is* it anyway?" Initially, they looked at the electron, which had just been discovered. Physicists noticed that when a charged particle (like the electron) moves, it makes a magnetic field. This magnetic field then pushes back on the particle, making it harder to get the particle moving faster. It's the same effect as if the electron had some kind of hard-to-push mass stuff to it, which gave physicists the first idea that mass and energy (in this case, the energy of the magnetic field) could be more than just two different things.

Then, Einstein stepped in with a clever argument that settled the debate.

At the time, Einstein had been preoccupied with the idea of *relativity*, the study of how the laws of physics apply to things that are moving relative to one another. It was known back then that nothing can move faster than the speed of light, and that this speed limit worked no matter how fast you were moving. If you were moving really fast, you would still see light moving at the speed of light. This fundamental limitation makes for some really strange effects when you consider how things look to someone standing on Earth and someone going really fast on a rocket ship.

For example, Einstein considered the case of a rock in space giving off heat. That heat will come off the rock in the form of infrared photons. If you are floating in space next to the rock, you might not notice anything strange. You would see photons coming off the rock and you would measure that the photons had a certain energy (as all photons do).

But if you were traveling past Earth on a speeding rocket ship, you would see something different. Einstein used the formulas of relativity to figure out that you would see the photons coming off the rock at a different frequency of light. This is an effect called the relativistic Doppler effect, which is similar to how, for example, a police siren sounds different if the police car is coming toward or away from you. In this case, though, the shift is a little stranger because of relativity rules (since you can't see the photon going faster or slower than the speed of light). The net effect is that you, in the spaceship, would measure the energy of the photons to be different than if you measured it when you're floating next to the rock. But since it's the same photons, something else must have changed.

According to Einstein, what also changed was the kinetic energy of the rock. But kinetic energy comes from the mass and the velocity of an object, and since the rock's velocity didn't change when it

gave off photons, Einstein concluded that its mass must have changed. In fact, he found that the mass of the rock changed by an amount equal to the energy of the photons, if you multiplied it by the speed of light squared. In other words, he found the following:

Energy of the photon = (Change in mass of the rock) x (speed of light)^2

What this means is that when a photon leaves the rock, it actually changes the mass of the rock. This change in mass is the same (if you multiply it by the speed of light squared) as the energy of the photon emitted. It seems that a little bit of the mass of the rock was transformed into energy, which then went off in the form of a photon (remember that photons don't have any mass; they are pure energy).

This was a pretty groundbreaking result, to say the least. It threw out thousands of years of human intuition that told us that mass and energy were totally different things. Instead, Einstein's equation says that the two things are related to each other and that you can somehow transform one to the other in the same way that you can walk into a currency store and trade dollars for euros.

At this point, you might be wondering: What does this mean? How exactly can something like mass, which has substance, be transformed into pure energy and vice versa?

Initially, you might think that a few of the rock atoms somehow disintegrated and became these photons. That would be one way for the mass of the whole rock to decrease. But actually, this is not what happens at all. The rock has the same number of atoms before and after the photon is created, but somehow, the mass of the rock decreases.

This is very strange to us because we're not used to the mass of things changing. If you have a metal weight on your desk, you don't expect that weight to be lighter or heavier just because you turn the air-conditioning on or off. A pound of sugar is a pound of sugar, regardless of whether or not you put it in the fridge, right?

To understand what's really going on here, we have to dig a little deeper into what it means for something to have mass. In particular, there are two important clues that will help us put this puzzle together.

Most of Your Mass Isn't Stuff

You probably think of yourself as being made of solid "stuff." After all, you are what you eat, and you eat stuff. Not lightning or sunbeams. And when you poke your finger against your arm, it feels pretty solid.

But actually, if you look closer and zoom into the bits that make you up, you'll see that there's really not much there at all. If you look at any particular atom in your body, a lot of it is empty space. Almost all of the mass of the atom is found in the nucleus, since the proton and neutron each weigh two thousand times as much as the electron. And as we saw in Is an Afterlife Possible?, even more fascinating is that when we crack open a proton or neutron, we see that they are actually made of "up" and "down" quarks: two up quarks and a down for the proton, and two down quarks and an up for the neutron.

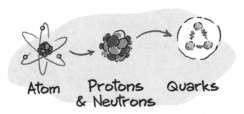

Atom   Protons   Quarks
        & Neutrons

So really, most of the mass in your body is in those groups of quarks. But what's really interesting is what happens when you separate those quarks.

If you measure the mass of the three quarks together (in the proton, for example), you will find that they have a mass of about 938 MeV/c^2 (one MeV/c^2 is about 1.7 x 10–30 kilograms).

But if you break open that proton and separate the three quarks, you will find that each up quark only has a mass of about 2 MeV/c^2 and that the down quark only has a mass of 4.8 MeV/c^2.

The quarks hardly have any mass by themselves! They each weigh less than 1 percent of the mass of the proton.

And yet when you put the quarks together, somehow their mass increases by a factor of one hundred. That's like putting three Lego pieces together and then finding that all of a sudden they weigh as much as *three hundred* Lego pieces. What's going on? Where does all this mass come from?

The surprising answer is that the mass comes from the energy that ties the quarks together.

You see, one amazing fact we've learned is that energy acts like mass. If you have a bit of energy in one spot—say, trapped in the bonds between two particles—that little bit of energy will be hard to push and pull in the same way that mass is hard to push and pull. If you separate the two particles and let the energy dissipate, then the particles will be easier to move around. In other words, energy itself has inertia.

And not only that, energy also feels gravity. Any little bit of trapped energy will also bend space and get attracted to other objects, just like things with mass do.

So in the case of the proton, its mass is the sum of the individual masses of the three quarks, *plus* the energy of the bonds that hold

them together (for quarks, it's the strong nuclear force that binds them).

This is true for all things in nature, not just protons. The mass of, say, a llama is equal to the mass of all its particles plus the energy needed to keep all those particles together (including the regular chemical bonds between the molecules). If you were to split the llama in two (sorry, llama), the sum of the masses of the two pieces would be less than the mass of the original llama.

And how do we figure out what the equivalent energy of that lost mass is? You guessed it: we use $E = mc^2$.

This is partly what $E = mc^2$ means: that mass is equivalent to energy. And it turns out that most of what we consider to be our mass (about 99 percent) is actually just energy.

The Other 1 Percent

What about the other 1 percent of us? That's still *stuff*, right? Actually, not so much.

In the last hundred years, we've also learned a lot about the nature of the mass of the fundamental particles. We've looked as closely as we can, and so far, particles like quarks and the electron don't look like they are made of smaller pieces. This tells us that their mass doesn't come from the energy of holding smaller bits together. So where does their mass come from?

The original idea from the 1880s was actually on the right track. Electrons are harder to move because of the magnetic field that they generate. But there's another field out there that also pushes back at them: the Higgs field. Filling the universe, this quantum field tugs at all of the matter particles, making them harder to move. This is where the mass of each particle comes from. It's each particle's interaction with the Higgs field. But that's only a partial explanation.

The full explanation is that the mass comes from the *energy* of the Higgs field. Some particles interact a lot with the energy stored in the Higgs field, which makes them harder to move. And some particles have weaker interactions, which makes them easier to move. In other words, the mass of each particle is really just the strength of its connection to the energy of the Higgs field.

And we can go a step further still. According to quantum theory, quarks and electrons themselves are nothing more than little ripples of energy in the quantum fields that permeate the universe. A particle is just a burst of energy, in the same way that a shout is a ripple in the air, or an ocean wave is a ripple in water. In other words, even the particles themselves are just energy!

A Massive Conclusion

Both of these clues—that most of the mass of an object is the energy of the bonds that hold that object together, and that even the mass of each particle is really just energy—lead us to an amazing, and somewhat shocking, conclusion: what we thought of as "mass" doesn't really exist. It's all just energy.

This is how the rock in space is able to lose mass when it radiates a photon. It doesn't lose mass because it transformed matter into energy. All matter is already energy. The rock just transformed energy from one form to another. In this case, it transformed the energy in the motion or vibration of the molecules into a photon.

So when you think of the rock in space, don't think of it as having mass *and* energy. Just think of it as one big blob of collected energy. Some of that energy is in the particles, some of it is in the bonds between the particles, and some of it is in the motion of the particles, but it's all just one pool of energy.

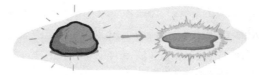

The reverse can also happen: if the rock absorbs a ray of sunlight and heats up, it adds to the pool of energy. And more energy means that the rock will be harder to move and that it will weigh more due to gravity. This means that hot rocks really *are* more massive than cold rocks. Of course, the difference is small: remember that to calculate the equivalent mass change, you have to divide the energy of the photon by the speed of light squared, which is a big number.

This is what $E = mc^2$ reveals: that mass is equivalent to energy. These days, physicists say that mass is a *form* of energy. This is because there are other forms of energy. For example, photons can have energy but not mass.

Just Do It

The famous formula does tell us that there's a deep connection between mass and energy. But it's not that mass is something that can be transformed into energy. What we've learned is that all mass is just energy. It's the energy of the particles of an object, either in their bonds with one another or in their interaction with the Higgs field.

The idea that energy has inertia or that it weighs something feels odd and counterintuitive, but that's only because we've been thinking about mass the wrong way for hundreds of years. There is no such thing as "stuff"; there's only energy and the impact it has on the shape of space (gravity) and how things move (inertia). Those are the two sides of Einstein's two-part relativity tango.

Fundamentally, this changes how we look at the universe. We no longer see the universe as filled with matter and energy. The entire universe is just energy, including us. In a real way, we are luminous beings made of energy.

Just don't expect to shoot lasers from your eyes anytime soon.

Where Is the Center of the Universe?

The center of anything is an important place.

For example, the center of your city is a landmark. It's where most of the action is, like where the best bakeries are and where important decisions are made. The very center of the city is also usually the oldest part, where the first loaf was baked and the first house was built.

YE OLDE BAKERY

The same is true, on a really large scale, for a lot of things in space. Our solar system has a center: the Sun! It's the first thing that formed out of the cloud of gas and dust that made us, and it's still the densest spot. It's also the best source of light and energy, and it's definitely the busiest place in the solar system: the lights never go out on the Sun. Even our galaxy has a center—a supermassive black hole with the mass of millions of stars whose gravity helps keep everything in place.

But centers are also important because they give you a sense of place. They help orient your location and give you a hint about where you are relative to everything else. Without knowing that, you might feel a little unmoored or lost, like being out at sea without a compass or getting stuck inside an IKEA store.

So what about the whole universe? Does it have a center, where everything started and where all the important universe business happens? And if it does, how close or far are we from it? Do we live near the center of the action or are we out in the middle of nowhere, cosmically speaking?

Let's take a look around and see if we can pinpoint the center of, well, *everything*. Who knows, we might even find some action when we get there.

What Can We See?

You can usually find the center of a city by looking at a map. Unfortunately, we don't have a map of the whole universe, because we can't see all of it. This is not because there's something blocking our view, or because the universe is too big, but because the speed of light is just too darn slow.

While light is pretty fast compared to competitive IKEA shoppers and airplanes, it's not infinitely fast. It takes time to cross the zillions of miles of space to bring us images of the distant reaches of the universe. And unfortunately, the universe is too young for us to see all of it. Physicists believe the universe started fourteen billion years ago, which puts a limit on the photons that we can see. If something is so far away that its light would take longer than fourteen billion years to reach us, then we can't see it. That means that the farthest thing we can see is something that sent light in our direction just after the universe started. Anything farther than that, and not enough time has passed for the light to get here, though it's on its way.

This volume of space that we can see is something we call the "observable universe." And since light travels at the same speed in all directions, this volume is a sphere centered around you (or more accurately, your eyeballs).

To be sure, the observable universe is huge. Because the universe is expanding, it's actually bigger than 14 billion light-years in every direction. Objects whose light is arriving now after 14 billion years are actually even farther away now because space got bigger. This expansion of space has stretched our field of view up to around 46.5 billion light-years, making the observable universe 93 billion light-years wide. And if we were looking for the center of the observable universe, the answer would be simple: it's you. We are each at the center of our own observable universe, since we all receive photons at a slightly different location.

And in fact, each person's observable universe is growing every year. Not just because space is still expanding but because as time passes, more photons are able to reach us, letting us see things that are farther and farther away.

But of course, the observable universe is not the same thing as the actual universe. Our limited view doesn't necessarily tell us whether the universe has a center. It could be that the observable universe is almost the same size as the actual universe, in which case we might start to get a sense of where the center is soon. Or it could be that the universe is much bigger than what we can see, and our little bubble of vision is off in a sad corner, missing all the fun.

Guess there's not much going on in the Universe.

Hints from the Structure of the Universe

Even though we can technically see to the edge of the observable universe, we've only just started to look around and figure out what's in our neighborhood. Only recently have we been able to build telescopes powerful enough to get a closer look at those far-away and dim galaxies.

And when we look around, the first thing we discover is that stars and galaxies are not evenly spread out across the universe, like chocolate chips in a well-baked muffin. Instead, they are arranged in large structures that gravity has managed to put together after fourteen billion years of patient work.

Our galaxy is part of a little cluster of neighborhood galaxies called the "Local Group." The galaxies orbit a common central point, swooshing around in space and occasionally running into one another. Our neighbor, Andromeda, will crash into our galaxy in about five billion years. There are other similar galactic clusters nearby, and together we all form a supercluster of galaxies that's many millions of light-years wide.

But superclusters like ours are not the biggest thing in the universe. In the last decades our telescopes have revealed that superclusters form even bigger structures: the walls of enormous bubbles that encase billions of cubic light-years of *nothing*. We are still putting together the whole picture, but as far as we know, those bubbles are the biggest structures in the universe.

Does that tell us where the center of the universe is? It would be great if the structure that we see told us something about where the center might be. Maybe we could see a pattern, like the way build-

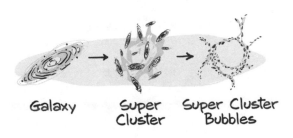

Galaxy   Super   Super Cluster
              Cluster   Bubbles

ings tend to get bigger as you get closer to downtown, or how galaxies get more crowded near their middle.

Unfortunately, not even these giant bubbles tell us much about where a center could be. They seem to keep going in every direction quite evenly. It doesn't get denser on any particular side or suggest any kind of pattern to find the center.

Hints from the Movements of Galaxies

Another way we might be able to find the center of the universe is to look at how all those galaxies and superclusters are moving. After all, we can tell where the center of the solar system is just by looking at the trajectories of all the planets. In the same way, you can trace the center of a galaxy by looking at the paths of all the stars in it.

It turns out that everything we see in the universe is also moving. In fact, we think stuff has been shooting through space since the very first moment: the Big Bang. Could the motion of all the things in the universe tell us where the center of the universe is?

Most people imagine the Big Bang as an explosion. They think that all the stuff in the universe was crunched into a tiny dot, and then it all exploded out through space. So if we look at the direction everything is headed and run the clock backward, would that

tell us where the center of the explosion was? Can we triangulate the Big Bang to find the center of the universe?

To try to figure this out, we have measured the velocity of many of the galaxies we can see. We do this by looking at the color of the light they shine at us. Just like a police siren sounds different moving toward you or away from you, the light from galaxies changes frequency if the galaxies are moving. Galaxies moving away from us look redder, while galaxies moving toward us look bluer.

And what do we see? We see that galaxies are indeed moving, and they're doing it at different speeds. But then we noticed something surprising: the motion of all the galaxies tells us that they're all moving away from . . . us!

Seriously, was it something I said?

Does that mean that *we* are at the center of the universe? Did the Big Bang happen *in this spot*, and now everything is flying away from it?

Not exactly. You see, the Big Bang wasn't actually an explosion; it was more of an *expansion* of space.

What's the difference? When a bomb explodes, it pushes everything away from the center. All of the bits of debris move away from a single point, and if you reverse their path, they would point back to the origin. That's why it's easy to tell where a bomb exploded. All you have to do is trace where all the debris came from.

But an expansion happens *at every point*, not from one center. It's more like a loaf of bread rising in the oven. It doesn't just grow from the center, pushing outward. The little air bubbles in every part of the dough all grow at the same time, puffing up the loaf evenly. And if you were inside the expanding loaf, you would see every part of the bread moving away from you, regardless of where you were. And that explains why we see things moving away from us in every direction: you would see that no matter where you are in an expanding universe.

THE UNIVERSE RISES

Unfortunately, that also means we can't use the expansion of the universe to tell where the center of everything is. Like an expanding loaf, all we know is that the universe is growing everywhere, so the center could be here or it could be anywhere.

And sadly, we also can't tell where the center is from the motion of the bubbles and superclusters. It would be great if they were all moving in orbit around a central spot, but so far it doesn't appear that they do.

Finding the Crust of the Universe

Does all of this mean that we will never find the center of the universe? Not necessarily.

Some of you might be thinking that just because a loaf of bread is expanding everywhere doesn't mean that it doesn't have a center. And you would be right. A loaf of bread is both expanding at every point *and* it can have a center. But it depends on the shape of the loaf.

One way to define a center is by geometry. For the bread, this is the point in the loaf where there's the same amount of bread in every direction. You can figure this out by tracing out where all the edges of the bread are (i.e., its crust) and then finding the middle point of all those edges.

BREAD JOKES NEVER
GET STALE

Could we find the center of the universe in the same way? Sure, but that depends on whether the universe even *has* a shape!

The problem is that we just don't know if the universe has a crust like a loaf of bread. We can't tell what's beyond the edge of the observable universe because we can't see that far. But there are a few possibilities.

The Universe Has a Blobby Shape

If the universe does have a shape, it *could* look like a loaf of bread, in which case it would have a center. It could be that this center is important, like maybe it contains some of the earliest matter that formed in the Big Bang, or maybe it technically occupies the spot where the rest of the universe came from. But it could also be that this center isn't very special. Maybe it's just the place that happens to be right in the middle. For example, take Oklahoma: it's right in the center of the United States, but few people would consider it especially important (sorry, Oklahoma).

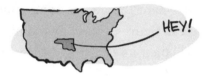

The Universe Is Infinite

It could also be that the universe continues to do its thing, filling space with bubbles of galaxy superclusters forever. Forever is a hard concept to grasp, but it means that you can travel in any particular direction and you would never run out of universe. This might sound strange, but a lot of physicists say that an infinite universe makes more sense that a finite one. And if it's true that the universe is infinite, it would mean something shocking: *the universe has no center.* If you define the center to be the point where there's the same amount of stuff in every direction, then every point in an infinite universe satisfies that definition, because there's infinite stuff in every direction.

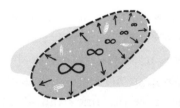

The Universe Has a Funny Shape

The final possibility is that the universe has a finite shape, but that shape can defy having a center. How is that possible? Well, it turns out that space can bend, so it doesn't always go in a straight line. This means that you can shape space in all kinds of interesting ways. For example, it could be that the universe curves back on itself, just like the surface of Earth curves around. If that's the case, then where is its center? Just like Earth's surface doesn't have a center (still not you, Oklahoma), the universe could also avoid having one. It could also be that the universe curves in a weird way, like in the shape of a donut. In that case, the universe would have a center, but that center wouldn't be inside the universe!

The Holey Center

While it doesn't seem likely that we'll ever travel far enough to personally check whether the universe has a crust, or whether it's infinite or shaped like a donut, it might still be possible to know which of these possibilities is true. By studying the nature of space and looking at its curvature around us, we might one day be able to infer what the overall shape of space is. It might tell us whether it goes on forever or loops around, or it might point us in the general direction of a geometric center.

The Central Point

Unfortunately, we don't currently know where the center of the universe is, and we might never know. We don't even know if the universe *has* a center!

But regardless of whether there's a center to the universe or not, there is a silver lining. We know for sure that the universe is expanding everywhere. And we also know that the Big Bang was not an explosion into empty space but an expansion of space itself. In a way, this tells us that everywhere is equally important, and that no place in the universe is more special than any other place. Like the loaf of bread, every point in the universe is the site of new space being created, which means that every point is the center of its own little universe.

For a physicist, that scenario feels more natural because the laws of physics shouldn't favor any one point over another. If there was a center, then physicists would ask, "Why that point?" and "Why not some other point?" It's much simpler to assume a democratic universe.

So in the end maybe we don't need to know where the center of the universe is. Each of us could just be content with being the center of our own observable universe, mooring ourselves to other people's view of the universe, and growing our awareness and perception of it as it continues to expand (perhaps infinitely) in all directions.

THE UNIVERSE: YOU
GOTTA LOAF IT.

PS: For bonus homework, go to Google Maps, search for "center of the universe," and then zoom out to see where it is.

Can We Turn
Mars into Earth?

E arth is pretty great, right? It offers incredible views, tasty street food, and good schools. As long as we take care of it, humans should be able to live comfortably on it for a long time. But is Earth the only planet that we can live in? Unfortunately, looking around our solar system, there aren't any other options that have the same luxury amenities, or even basic things like reasonable temperatures, a breathable atmosphere, or liquid water on the surface.

Even if we found another planet like Earth out there, traveling to it would take dozens, hundreds, or thousands of years, unless we invent warp drives or find a way to manipulate wormholes. Instead, what if we found a fixer-upper closer to home? One that might need some work and a coat of paint, but that we could travel to without spending decades crammed inside a smelly colony ship?

Well, look no further than the planet that is literally next door: Mars! It needs a bit of work and updated bathroom fixtures, but it has real potential. And it scores very highly in the three most important categories: location, location, and location.

What would it take to give Mars a makeover? And can we make it as nice as Earth?

Living on Mars

When we say that we'd like to make Mars livable, we mean that we'd like it to be as similar to Earth as possible. We could, in theory, build space stations to live there, where you have to wear a fancy suit to go outside. You could even build massive domes to enclose cities and stay indoors the whole time. But what kind of life would that be?

To really call a place home, we want to be able to roam free, breathe fresh air in green parks, and enjoy the land. We don't want to have to wear space suits to go for a walk, or slather on 1000 SPF lotion to avoid cosmic radiation.

The problem is that Mars isn't quite in move-in condition. To make Mars more like Earth, we'd need to change several things about it that make it a not-so-nice place to live in right now:

- ✦ There's no liquid water on the surface.

- ✦ It's very cold there (think Antarctica all year round).

- ✦ There's no breathable atmosphere.

- ✦ The surface is bombarded by harmful cosmic rays.

Let's tackle each of these one at a time.

Water, Water Everywhere

Everyone knows that water is connected to life. Not only does all life (as we know it) need water to survive but we think that life started in the water. When we look around the solar system for possibilities of alien life, one of the first questions we ask is: Where is there liquid water? So far, Earth is the only place in the solar system where we have found liquid water on the surface. And that's what we want: easily accessible liquid water, preferably in beautiful lakes and running streams.

Of course, "liquid" is the key term here, because water as a molecule is not actually that rare in the solar system. In fact, Uranus and Neptune are called "ice giants" because they have so much solid water. The dwarf planet Ceres is estimated to be half ice, and a lot of the rocks in the asteroid belt are basically huge dirty snowballs. In fact, scientists think that much of the water on Earth came from the far reaches of the solar system. A young, hot Earth boiled away a lot of its original water into space, but its water was later replenished by impacts from comets and other icy space rocks. That's right, our oceans are *filled* with melted cosmic snowballs. The next time you drink a glass of water, remember you're enjoying a cool, refreshing glass of melted comet.

NEW BUSINESS IDEA

Mars definitely doesn't have any surface oceans, but it still has plenty of frozen water aboveground and liquid water deep underground. Mars, like Earth, is colder at its north and south poles than at its equator. And its poles are covered in ice, just like those on Earth. Lots of ice. So much ice that if you could melt all of it, Mars would be covered in water to a depth of one hundred feet. That's plenty of water for future humans living there to drink, swim, and build waterslides for their theme parks.

If we want oceans and rivers in our new home, all we have to do is melt it and keep it melted. But that's tricky because Mars is very, very cold on its surface, and it has a very thin atmosphere. Any liquid water out in the open is most likely going to freeze, or boil into vapor in the vacuum of space.

The good news is that if we can figure out a way to heat up Mars, and also find a way to give it an atmosphere, then Mars can have liquid lakes and oceans and be one step closer to resembling our dear planet Earth.

Making Mars Toasty

From the look of it, you might imagine that the surface of Mars is toasty and warm. It shines red, after all, and it mostly looks like a desert. But Mars is actually very cold. Its red color comes from all the oxidized iron in its soil. The average temperature on Mars is −81°F, which is much colder than the temperature on Earth's South Pole.

If we want to turn up the thermostat on Mars and make it a cozier place to live, we need to think about what gives a planet its temperature. A planet's surface temperature is mainly set by two basic things:

(a) How much heat it gets from the Sun

(b) How much of that heat it holds on to

Most of the heat in the solar system comes from the Sun, so the amount of heat a planet gets depends on where it is in the solar system. The closer a planet is to the Sun, the more heat it gets. Mars gets a decent amount of heat, since it is the fourth-closest planet to the Sun. But it doesn't get as much as Earth, which is one planet closer.

One potential solution is to change the distance between Mars and the Sun. We could build giant planet-size rockets and strap them to Mars to steer it to a closer orbit. A cheaper but more dangerous idea would be to use another heavy rock as a gravitational tugboat. If we could steal a big asteroid and put it in orbit near Mars, the gravitational effects could pull Mars in the right direction. Assuming, of course, that we don't slam that asteroid into the planet.

If that sounds a bit crazy, then maybe we should consider other promising solutions. For example, we could increase the temperature of Mars by helping it keep more of the energy it gets from the Sun. Planets don't wear puffy down vests or parkas to stay warm, but they do have atmospheres. Atmospheres aren't just great for breathing and giving you beautiful sunsets; they can also act like jackets to planets, thanks to the greenhouse effect.

ATMOSPHERES: ALWAYS
A HOT ACCESSORY

When light from the Sun hits a planet, it heats up the rocks and mountains and all the things on the planet surface. When these things warm up, they glow in the infrared.* Normally, this energy would just radiate out into space and get lost. But if you have an atmosphere, that radiation can get trapped inside. The key is to have carbon dioxide (CO_2) in your atmosphere.

CO_2 works like a one-way mirror because it only absorbs a particular kind of light: the infrared kind. Visible light from the Sun

---

* That's just because Earth and Mars are cooler than the Sun. Everything in the universe glows at a wavelength determined by its temperature. The Sun glows in the visible spectrum, and planets like Earth glow in the infrared.

passes through the CO_2 on its way in, but when that light gets reflected back as infrared, it gets blocked by the layer of CO_2, trapping the energy inside and warming up the planet. Of course, you can imagine how having too much CO_2 can cause your planet to overheat, too.

Mars does have an atmosphere, and most of it (about 95 percent) is CO_2. Sadly, though, Mars's atmosphere is pretty thin. In terms of pressure, Mars's atmosphere is less than 1/100th the pressure of Earth's. So most of the sunlight that falls on Mars just gets radiated back out into space.

We could bring Mars up to a warmer temperature by engineering a massive atmospheric makeover and increasing the amount of CO_2 in the atmosphere. But to get that full greenhouse effect, Mars would actually need more CO_2 than we have in our atmosphere on Earth, since Mars gets less sunlight than we do. So where do we get more CO_2?

Until recently, most of the CO_2 on Earth came from volcanic eruptions. But Mars doesn't have any active volcanoes that can spew out CO_2. The interior of Mars is cold and hard, without the flowing rivers of molten lava that powers volcanoes. Scientists think that millions of years ago it was a different story, and Mars was hot and melty on the inside. But Mars is smaller than Earth—about half of its diameter—so it cooled and hardened faster than Earth, like a smaller cup of coffee on a winter morning.

One little bit of good news is that Mars already has a small source of CO_2 that we could use. Those layers of ice at the poles are not all made of frozen water. A lot of that ice is actually frozen CO_2. Yes! That's just what we need. If we could somehow melt the poles, we would unlock plenty of water, and release a little bit of CO_2 to help keep it warm.

Unfortunately, even if you released *all* of the CO_2 on the poles, you would only get about 1/50th of the CO_2 you'd need to keep the planet warm and toasty.

Are there other sources of CO_2 we could find? Actually, the solar system has plenty of frozen CO_2 in asteroids and comets. One potential solution is to send spaceships to nudge some comets and get them to smash into the surface of Mars.* It would take a *lot* of comets, probably thousands or millions of them.

Before you start building your fleet of comet-steering spaceships, there's another issue. The amount of CO_2 you need to keep Mars warm would unfortunately also make the air poisonous for humans to breathe. We can tolerate a bit of CO_2 in our lungs, but if

---

* Probably best to do that *before* we send people over.

you have too much, you start to feel drowsy, get headaches, sustain brain damage, and eventually die. Unfortunately, there's no happy ending to blanketing Mars with more CO_2.

There is another way to heat up the planet, though. We could capture more of the Sun's rays and direct them to the surface of Mars by using huge space mirrors. How huge? To gather enough light to warm up Mars, we would need space mirrors *the size of Mars*. That's not a small project. But it would provide the heat we need to release the CO_2 and water at the poles and make Mars both warmer and wetter.

Oh Yeah, Oxygen

If we manage to get the air temperature just right and melt the ice in the Martian polar regions to make new rivers and lakes, we still have a lot of work to do to make Mars into a capable replacement for Earth. We need to be able to breathe the air. Specifically, we need oxygen! Nobody wants to strap on a breathing mask every time they go for a picnic or borrow a cup of flour from the neighbors.

While oxygen is very common in the solar system, the kind that we need to breathe is surprisingly hard to find. Human lungs need the oxygen molecule O_2, which is a pair of oxygen atoms tied together. There is plenty of oxygen in the universe; it's one of the lighter elements, so it's made in vast quantities by the fusion at the heart of stars. But oxygen is a very friendly atom, and likes to bond with basically everything else around it. On Mars, there's oxygen in water (H_2O) and in carbon dioxide (CO_2), but almost no pure O_2.

On Earth, our air is around one-fifth O_2. In our case, it wasn't made geologically but as the byproduct of early life. Most of the original O_2 on Earth was produced by tiny little organisms in the ocean. These early bacteria were doing photosynthesis long before there were even plants. Around 2.5 billion years ago, those little organisms drank sunlight and water and CO_2 and burped out O_2 in the process. There wasn't any oxygen-breathing life at the time, so the amount of oxygen steadily increased for millions of years (maybe even a billion). Later, these microorganisms were incorpo-

rated into plants, which continue to pump out the O_2 we need to breathe.

Could we somehow make this happen on Mars? It sounds promising: a little biological machine that will use the sunshine and the newly melted water and CO_2-rich atmosphere to create O_2 for us. Even better, these organisms multiply on their own, so we just have to plant a few batches of them on Mars and they will make more of themselves. It's like crowdsourcing on a whole new level, and we can pay for it with sunshine.

As usual, though, there's a catch. On Earth, this process took a long time, maybe a billion years. That wasn't inconvenient for us because it started long before humans existed. If we had kicked off this project on Mars a billion years ago, it would be ready for us just about now. Short of building a time machine, are we doomed to wait a billion years for Mars to have a breathable atmosphere? Microbiologists have lots of tricks to make bacteria grow faster and work harder (and take shorter lunch breaks). But it's still a very big job for tiny little organisms, and even the sped-up version of this process will likely take many thousands or even millions of years.

How else can we fill Mars with oxygen? One solution is to build oxygen factories, which produce O_2 chemically instead of biologically. It might sound like science fiction, but actually, an early prototype of this device is on its way to Mars right now, as part of the Mars 2020 mission. NASA has made these machines mainly to make O_2 as rocket fuel for missions that will return samples from Mars, but in principle the same concept could be used to make breathable oxygen.

It's like pollution, but good.

Magnetic Field

Once you have spent billions of dollars producing a nice atmosphere (or enslaved zillions of bacteria to do it for you), you probably want it to stick around. It would be an epic fail if your atmosphere just got blown away, like cosmic dandelion fluff.

If you're thinking that's impossible because there's no wind in space to blow the atmosphere away, then let us introduce you to a whole different kind of wind. The "solar wind" is made of fast-moving particles from the Sun. It's mostly protons and electrons made in the same reactions that make all that beautiful sunlight. And there are also particles coming from deep space called "cosmic rays." None of these particles are harmless. Quite the opposite, actually, as they are fairly deadly. Astronauts in space have to wear heavy shielding to protect themselves from this harmful radiation. This stream of high-speed, tiny bullets will strip any planet of its atmosphere given enough time.

Thankfully, we have an awesome planetary protection system here on Earth: our magnetic field. When electrons or protons hit a magnetic field, they get deflected. Our magnetic field deflects many of the Sun's harmful particles, making them miss Earth or spiral up to the poles, where they create the dazzling northern and southern

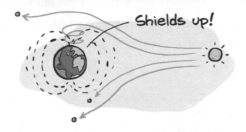

Shields up!

lights. Without our magnetic field, we would be blasted by harmful solar radiation, which would also strip away our atmosphere.

Unfortunately, Mars doesn't have a planetary magnetic field like Earth's. On Earth, our magnetic field is created by the rivers of molten metal flowing within our planet. Mars, however, is a smaller planet, and it has cooled earlier than Earth, which froze its inner core and killed its magnetic field. Without that magnetic field, anyone on the surface of Mars will need serious protection from radiation—thick suits lined with lead. That's not something you want to have to put on every time you go outside to kick a ball around with your Martian kids ("Mom, I have to pee . . ."). And without that magnetic field, any atmosphere you create will eventually get blown away. This is a bigger problem on Mars than on Earth because Mars has weaker gravity, so it's harder to keep the air molecules on its surface.

We could potentially restart Mars's magnetic field by heating the core and getting those metals flowing again, but jump-starting a whole planet is engineering on a scale even we can't imagine.

There is hope, though. Maybe we can build something that does the same job. Engineers at NASA came up with the clever idea of building an artificial magnetic shield, but instead of trying to wrap it around the whole planet, they suggest a smaller shield that is close to the Sun. Being close to the Sun lets the shield cast a larger magnetic "shadow." The shield would sit in space between the Sun and Mars, deflecting much of the solar wind to protect the atmosphere from getting blown away.

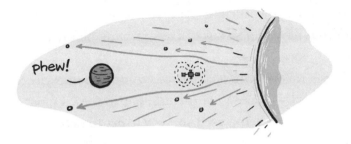

Other Homes?

This might all strike you as a lot of work. To summarize, turning Mars into a planet like Earth would require:

+ A massive set of solar mirrors to focus sunlight and warm the planet

+ A vast array of planetary factories to generate oxygen for us to breathe

+ A space-based magnetic shield to protect the new Martians and their atmosphere from solar radiation

Perhaps you are thinking that Venus or the moon, which are also close by, would be better candidates.

Unfortunately, Venus has the opposite problems as Mars. The surface is blanketed with huge amounts of CO_2, which poison the air and trap heat. And since Venus is closer to the Sun than Earth, it gets more sunlight, which raises its surface temperature to a toasty 460°F. All that trapped energy also makes the atmospheric pressure at the planet surface so great that spacecraft landers we've sent to Venus have only survived a few minutes before being crushed to bits.

Of course, this hasn't stopped some wild-thinking scientists from proposing wacky ideas: What if you scooped the CO_2 out of Venus (using giant spoons?) and used space mirrors to deflect some of the sunlight? Would that make Venus habitable? Others have suggested building cloud cities that would float fifty kilometers above the surface of Venus. At that level, the temperature and pressure are actually similar to those of Earth. Sadly, those clouds are made of sulfuric acid, which makes writing the ad copy for the real estate brochures a little tricky ("Live on Venus! The views will take your breath away . . . literally!").

MY GENIUS IDEA

The moon is even closer, but frankly, it's just not big enough. Its mass is about 1 percent of the mass of Earth, which gives it such weak gravity that it can't hold on to any atmosphere. The individual particles of air would routinely shoot off into space, so even if we imported the ingredients from Earth, it would all disappear within a hundred years.

So in the neighborhood, Mars really is the best bet.

Should We Move?

Mars might be our best prospect for a second home world, but it's definitely a serious renovation project. Making Mars habitable would probably cost trillions and trillions of dollars, and might take thousands of years. And that's the *initial* estimate. Contractors always find a way to charge you extras once they get started.

Of course, it all depends on how motivated we are to move. Perhaps we need to leave Earth because a giant asteroid is coming to hit us. Or perhaps we ruin our climate to the point where Earth will become even more uninhabitable than Mars in the future. Given the right incentive, building giant solar mirror arrays and huge oxygen factories might be our best option. And look at it this way: the surface area of Mars is about 56 million square miles (or 1,559,000,000,000,000 square feet). So if we end up spending trillions of dollars to make Mars habitable, it would still be cheaper than buying real estate in California.

Can We Build a Warp Drive?

T he universe is achingly vast, and filled with fascinating places we'd love to explore. Unfortunately, it all seems to be out
of our reach.

Arriving in
46 billion years

As we learned in a previous chapter, even if we could get a spaceship up to some reasonable fraction of the speed of light, it would
take hundreds of thousands of years just to get to the other side of
our galaxy, not to mention visit other galaxies (millions of years
away) or even to push beyond the observable universe (hundreds of
billions of years away).

And that is a definite limit. Few rules in the laws of physics are
as hard and unbendable as the fact that nothing can move through

space faster than the speed of light. This limit is based on our understanding of Einstein's theory of special relativity, which has been tested and probed and verified out the wazoo (and into the wazoo; really, we've tried *everything*).

It seems that the only way we'll reach the far edges of the universe is if we become a spacefaring civilization, hopping slowly from planet to planet over countless generations for millions or billions of years.

And yet it doesn't seem like that should be the case. We are conditioned by movies and books to think that the universe should be within our reach. That with the right technology, you can build vast space empires or explore other galaxies. You just hop on your spaceship, hit a button, and *whoosh*: the stars streak in front of you and light and energy swirl around you as you slip into "hyperspace," and then *boom*, you arrive millions of light-years away.

All you need is . . . a warp drive.

But what *is* a "warp drive"? Is it something that's totally in the world of fiction, or is it something that real physicists have considered? Is it possible to break the speed limit of the universe, the one that scientists cherish so much? Let's hit the button and see if we can warp our way to an answer.

Making Fiction Real

A lot of technological advances seem to happen in the following way:

Step #1: Science fiction author invents a new gadget, hand-waves the science.

Step #2: Physicists figure out how to make the gadget theoretically plausible, hand-wave how to build it.

Step #3: Engineers figure out how to build it, hand-wave how much it would cost.

Step #4: Yadda, yadda, yadda, it's now in your smartphone.

For warp drives, science fiction authors have done a solid job on Step #1, imagining portable warp drives that can take you to the stars. Now it's time for physicists to do their bit.

At first glance, you might think that physicists would say no. After all, a warp drive seems to break the one rule they seem pretty adamant about: getting somewhere faster than the speed of light. On this, physics will not wiggle or budge. *But,* if there's one thing most teenagers have learned, it's that if at first you don't get the answer you like, try asking a different question!

For example, if you ask the question, "Can we build spaceships that can travel through space faster than the speed of light?" the answer is a hard no. But if instead you ask the question, "Can we build spaceships that can get to places faster than light could have traveled there?" then you might notice physicists squirming a little before finally admitting, "Maybe." And every teenager knows that "maybe" is code for "I want to say no, but I need to check with your other parent."

The key difference in the two questions is the phrase "travel through space." If you read the fine print of special relativity, you learn that the speed limit applies to things moving *through* space. Now, that doesn't seem like it offers much of an opening, because doesn't everything move through space? The answer is yes, but the loophole is that space is . . . malleable.

"Conditions may vary. The Universe is not responsible for any damages due to irresponsible use of space..."

Who needs a lawyer when you have a physicist!

There are three general ways in which we can see a warp drive being feasible from a physics point of view:

✦ A hyperspace warp drive

✦ A wormhole-powered warp drive

✦ A space-bending warp drive

Let's get into each of these ideas and whether they are theoretically sound or even likely.

A Hyperspace (or Subspace, or Superspace) Warp Drive

In a lot of science fiction, the loophole for making a warp drive work is to leave our normal space (where the speed limit of the universe applies) and enter some other kind of space. Presumably, you can either go faster than light in this space or this space somehow connects the place where you are to the place where you are trying to go. Once you've traveled for some time in this hyperspace, you just slide back into normal space.

This approach works in fiction, allowing the characters and the story to span a whole galaxy without spending thousands of years

sitting in a spaceship. But does this have a basis in real physics? Is there another kind of space that is parallel to our universe that we can somehow go into and out of?

One common idea that sometimes gets attached to this concept is the idea of "extra dimensions." We know that our space has three separate directions of possible motion: you can call them x, y, and z, but those are just arbitrary names. Some physicists suspect that there could be more ways that you could move, extra dimensions of space. It's hard to think about how that would work or where they would be, but it comes up often in string theory and other creative theories about gravity. According to such theories, these extra dimensions are not like ours: they curl around themselves and have different rules for how particles move through them.

That seems a lot like what we are looking for, right? Different parts of space with new sets of rules. Unfortunately, it's not as helpful as it might sound. These additional dimensions, if they exist, are not a different kind of space that's parallel to our space. They are just an extension of the space we have. They don't let you leave the space that you're in right now; they just give you more ways for your particles to wiggle or shimmy. It's like adding another line to your mailing address. It more accurately says where you are, but it doesn't provide your mail delivery person any shortcuts to bring you your mail faster.

There is one real physics theory that closely matches this idea of hyperspace: the multiverse. This is the idea that there could be other universes out there, either alternate versions of ours (split during quantum events) or other pockets of space with different laws of physics or different initial conditions.

If there are other universes, could they let us skip around our universe? Only if they were smaller or had higher top speeds than ours, and somehow connected to our universe in several different places. In that way, you could potentially jump into that universe, travel a short distance, and then connect back to our universe at a point that is really far away from where you started. And hey, maybe that other universe does look like a swirling tunnel of light and energy.

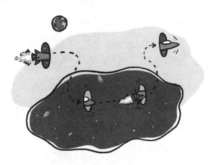

Sadly, though, the idea of the multiverse is still extremely theoretical. We don't have any reason to think it actually exists, other than to explain some weird quirks about our universe. And even if other universes do exist, physicists think the very thing that makes them appealing—their distinct rules of physics or alternate quantum variations—might also make it impossible for our universe to interact with them. So the most likely scenario is that we can never connect or travel between the different universes.

Wormhole Warp Drives

There are bizarre corners of our universe where space is bent and twisted beyond recognition, like no place we have ever known. The most famous members of this mysterious category are black holes, which are definitely not on the list of places we recommend you visit, since they are hard to survive and impossible to return from.

But there is a strange theoretical space-fold that might let you travel faster than light to a distant star: wormholes.

Wormholes are everywhere in science fiction. Writers use them as a shortcut between distant locations, to open portals to neighboring galaxies, to build exotic houses where every room is on a different planet, or to link planets into a galactic empire. In this way, you might imagine a wormhole as the basis of a warp drive: when you hit the button, you are opening and going through a wormhole that connects you to somewhere else in space.

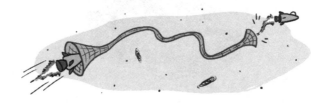

At first glance, wormholes seem totally impossible. Wouldn't it count as faster-than-light travel, which is a big no-no according to physics? It's certainly true that travel from point A to point B is limited by the speed of light . . . but only if you go through all of the space between them.

While physics can't bend the rules, it turns out that the rules themselves allow for some bending of space and some strange connections. When you think of space, you probably imagine it as a flat backdrop for the action of the universe. But space is much more interesting than that, and it can have all sorts of interesting shapes and be connected in all sorts of ways. Space is actually part of the action, not just the backdrop, because it responds to the matter and energy inside of it. Matter and energy tell space how to bend, and space tells matter how to move. It's like a cosmic tango.

If it's totally empty, space is boring and simple. But if you plop a big fat star in the middle of it, the star will bend the space. That means that it changes the shape of space, and gets matter to find new, curved paths along it. That's why photons bend around massive objects even though they have no mass. They are just following the curvature of bent space. Physics tells us that space might be able to have any smoothly varying shape. And one of those shapes is a wormhole, a strange deformation of space that links two points that are far away from each other.

Wormholes actually have a close relationship with black holes. One way to make a wormhole is to connect two black holes by their singularities, which are the points of infinite density at the heart of each black hole. If the two black holes are far apart, then the wormhole is like a shortcut through space, creating a connection between the two points.

But this kind of wormhole doesn't help us at all. Why? Because even if you survived going into the first black hole (a tricky proposition in itself, as we've discussed) and traveled to the other side of the wormhole, you'd still be trapped inside the other black hole! You might have traveled to another part of space faster than light, but you are never leaving that point ever again.

The kind of wormhole that would be helpful for a warp drive is one that lets you escape out the other side. The only way to do that

is to make a wormhole that connects a black hole to a "white hole." As we mentioned in an earlier chapter, white holes are theoretical objects, predicted by general relativity, that are the opposite of black holes. In a white hole, things can escape, but never enter. Think of the white hole as the exit point of the wormhole.

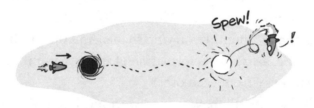

There are, obviously, a couple of problems with using this kind of wormhole for a warp drive.

First, it's a one-way connection. You might be able to fall into the black hole, pass through the wormhole, and then come out the white hole, but you can't go the other way. If you've figured out how to build wormholes and move their ends around, that might not be a problem for you, since you can just make another one to go back.

Second, it might be hard to survive the whole experience. Going into a black hole is no easy matter. Even if you pick a large black hole to go into to avoid getting ripped to shreds by its gravitational tidal forces, you still have to survive the trip to the middle of the black hole. And how do you squeeze yourself through a singularity?

For this, physics does have a cool answer: pick a spinning black hole. We prefer this kind of black hole because the center of it is not a tiny point but a spinning ring. Why is that? Things that fall into the black hole were probably spinning around it first, in the accretion disk. When they enter, that angular momentum can't just disappear. And since a point singularity has no size, it can't rotate and so it can't have any angular momentum. That's why a black hole with angular momentum has a ring at the center! And if it's connected to a white hole, then in principle you could pass through that ring and into the white hole.

A wormhole is also hard to keep open. The theory predicts that wormholes are prone to collapsing. The singularity ring at the center tends to pinch off and form two separate black holes with two

Doh!

singularities. You definitely don't want to be in the middle when that happens.

The final problem with using a wormhole for a warp drive is that, so far, it's all very theoretical. Nobody has ever seen evidence that they actually exist. All of these fun ideas depend on general relativity being correct (and, so far, it has passed every experimental test). But we don't know whether it's right about super-extreme scenarios, like the center of black holes, where quantum physics can't be ignored.

We know black holes exist (we've seen those), but wormholes and white holes are still just an idea at this point. We don't even know how to make them. Nobody has yet found the recipe to make a wormhole, much less how to specify which points in space it connects. Think about it: your spaceship would need to have the ability to create a specific kind of black hole, and then somehow connect it to a white hole a long distance away.

Still, if you were to find a wormhole, or figure out how to get the

universe to make one on command, they could potentially be used to warp-speed your way to the other side of the universe.

Space-Bending Warp Drives

So if hyperspace doesn't really exist, and wormholes end up being too dangerous to go into, are there any other clever physics loopholes we can use to make a warp drive? As it turns out, the answer is yes.

Space is much more interesting than we first imagined. It isn't "nothing," but rather a thing that can wiggle (as gravitational waves), bend (which is what gravity is), and expand (as we've seen with dark energy and the expansion of the universe). It seems that space can be stretched or compressed in response to the mass and energy around it.

So what if instead of traveling *through* 4.2 light-years of space like a galactic newbie, we squeezed the space between here and the point where we want to go? And what if, at the same time, we expanded the space behind us?

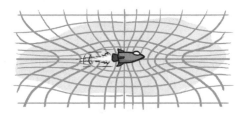

The idea is to reduce the amount of space that you have to travel through. You could squeeze the space in front of you, cross it, and then expand the space behind you so that it goes back to normal. For example, each step would go something like this: you squeeze the 1,000 kilometers of space in front of you down to a tenth of a nanometer; then you move that tenth of a nanometer; and then you expand the space behind you back to its original 1,000 kilometers. The end result is that you only moved 0.1 nanometers, but you actually traversed 1,000 kilometers. If you can do this continually, your spaceship would sit in a kind of inverted warp bubble, moving you forward at an incredible rate. For you inside the inverted bubble, 4.2 light-years would become 4.2 *kilometers* that you have to cross. Then when you get there, you pop the ship out of the bubble, and presto, you have arrived!

It's a little like taking a moving walkway rather than actually walking. Physics is very strict about how quickly you can move along the walkway, but there's no speed limit to how fast the walkway itself can move. In the same way, physics has no limit to how quickly space can stretch or compress or move relative to itself.

But how do you *shrink* or *expand* space? What does that even mean?

Making space shrink or bend is not actually terribly tricky. You are doing it right now. And every time you visit the dessert bar and gain weight, you get better at it. Everything that has mass changes the shape of space. It's why Earth orbits the Sun: because the Sun's enormous mass has bent the shape of space like a bowling ball on a trampoline. The bending is intrinsic, changing the relative distance between bits of space-time.

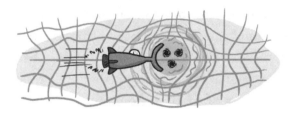

Unfortunately, while physicists know that a warp bubble satisfies the equations of general relativity, they don't know how to arrange matter and energy to make one. It's like having the idea for a complicated dessert but no clue what the recipe is for baking it.

The trickiest part is that the back half of the warp bubble has to *expand* space. We know that mass and energy can compress space, but how do you expand it? All of the space in the universe is currently expanding, as it did quickly in the first few moments after the Big Bang, and that expansion is speeding up. We say that it's due to dark energy, but that doesn't mean we know what dark energy is. It's actually the other way around: "dark energy" is simply the term we use to describe the increasing expansion of the universe. We don't actually know what is causing it.

To expand space artificially, physicists have proposed another crazy idea: If you can shrink space with positive mass, can you expand space using *negative mass*?

Negative mass? What does that mean? As far as you know, everything around you has either zero mass (photons) or positive mass (you, matter, bananas). That's why we say that gravity is a purely attractive force. Unlike magnetism, which can attract (fridge magnets) or repel (maglev trains), gravity only seems to attract—because we've only seen positive mass.

Is negative mass possible? It's theoretically possible, but to date nobody has ever seen any matter with negative mass. It would be really weird stuff that acts in funny ways. Positive mass attracts, so if we put a blob of it next to a blob of negative mass, the negative mass would push away the positive mass, but the positive mass would pull on the negative mass. Like a teen soap opera where you never know who's chasing who, it gets confusing quickly.

Now, supposing that we figure out a way to make negative mass, could we then actually make a warp drive like this work? Sadly again, there are other limitations. Expanding and contracting space isn't something that comes cheap. It requires energy.

Physicists first estimated that the amount of matter or energy needed to bend the space in front of a warp drive was larger than all of the stuff in the universe. That's clearly not going to work. A bit of tweaking on the calculations brought the estimate down to requiring the energy equivalent of the entire mass of the planet Jupiter. A gas tank that big would probably make your spaceship really hard to parallel park when you arrive at that other galaxy.

There's been some talk of reducing this further, to reasonable levels, such as the energy equivalent of a one-ton mass. But so far, this is still in the "physicists discussing this in the break room" level of scientific research. No one has actually built or tested a space-compressing machine yet, so this is still pretty far off in the future.

Speaking of negative mass...

WARP DRIVE SYMPOSIUM

A Warped Answer

As much as we'd love to find a loophole to the speed limit of the universe and conquer the stars, it seems that the idea of a warp drive is still very much in the realm of fictional galactic space operas. But as always, it's good to remember that the universe is unpredictable, and that human progress and ingenuity are still on the rise. Maybe one day we will figure out the details for creating black and white holes and connecting them across space. Or maybe one day we will discover negative mass and new ways to harness energy to make devices that let us squeeze into a warp bubble and zip our way to other galaxies.

True, that's a lot of maybes. But if you ask your other parent, they might just let you get away with it.

When Will the Sun Burn Out?

O ur sunny days are numbered.

From a distant ninety-three million miles away, the Sun seems to us like a strong, steady presence. It rises every day, without fail, and showers us with a constant stream of life-giving energy rays. But physicists have a very different view of the Sun.

Uh-oh.

To a physicist, the Sun is a constantly exploding nuclear bomb. This turbulent process releases vast amounts of energy, kept in check only by the sheer force of the Sun's gravity. The next time you enjoy a sunny afternoon, remember that you are toasting your toes to the light of a nuclear explosion. But physicists also know that underneath this incredibly tumultuous phenomenon there are mechanisms conspiring to end it, and that there's an internal clock steadily ticking down to zero. The physics of the Sun reveal that its days of shining brightly will someday end.

Is this going to happen anytime soon, or do we have billions of years to plan for it? Let's find out exactly how many sunny days we have left.

A Star's Birth

(5 billion years ago; Sun's age: 0)

To understand why and when the Sun will eventually die, we first have to go back to its beginnings.

The Sun wasn't born in some fiery, dramatic event. There wasn't even a little bang. Instead, there was a gradual accumulation of gas and dust. Most of the gas was plain old hydrogen, which has been the most common element in the universe since there *was* a universe. But there were other, heavier elements, too: leftover bits from nearby stars that had already lived and died by the time our Sun came around.

These vast, swirling clouds were slowly gathered together by gravity, which is the weakest (yet most persistent) force in the universe. But the gas and dust particles in these hot and swirling clouds were moving too fast to be completely held together by gravity, and they resisted forming dense clumps.

Scientists aren't sure what eventually triggered the formation of

TOO HOT TO CLUMP

our Sun. It could be that magnetic fields helped trap the particles and channeled them close together. Or it could be that some external event, like a shock wave from a nearby supernova, pushed the gas particles together tightly. Or maybe it was just time: eventually the gas cloud cooled, and the slower-moving particles started to fall toward the center.

Whatever the cause, enough stuff eventually clumped together to start a runaway process. Gas and dust gathered in one place, which led to stronger gravity, which attracted more gas and dust, which led to more gravity, and so on. Eventually, enough gas and dust had gathered in one place to form the beginnings of a star. And that's when things really started to heat up.

A STAR IS BORN

Fusion Pushes Back

(4.9 billion years ago; Sun's age: 0.1 billion years old)

After about one hundred thousand years, gravity had done its job of pulling together a huge cloud of mostly hydrogen. At first, the individual molecules resisted. They don't enjoy being crowded that close together because the positive charges of their protons repel each other. Getting two protons close together is like trying to put a cat into a bucket of water: you have to really want it. Fortunately, gravity never gives up. Over time, the enormous accumulated mass kept pushing and pushing the protons together until something finally snapped.

If protons come close enough together, they overcome their repulsion and start to *attract* each other. That's because a different force starts to act: the strong nuclear force. This may be the one thing in particle physics that is well named, because the strong force is, well, *strong*. Over long distances, it's not very powerful, but at short distances, it's much stronger than the electrical repulsion that keeps protons apart. Once this strong force brings the protons together, something incredible happens: fusion.

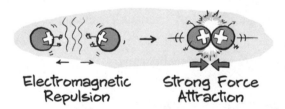

Electromagnetic
Repulsion

Strong Force
Attraction

The nuclei of the two hydrogen atoms stick together and, after a few more steps, eventually form a new element, helium. People tried for centuries to convert one kind of element into another (usually lead into gold) and failed for so long that the entire endeavor, known as "alchemy," was written off as crackpottery. It turns out that it's totally possible, but only under special conditions, like being at the center of the Sun.*

The amazing thing about fusing hydrogen into helium is that it releases a lot of energy. The helium that is made actually has less mass than the original hydrogen atoms, and the extra mass gets converted into energy, which is then carried off by neutrinos and photons. If you're confused by how making a bond can release energy, just think of the opposite case: it usually *takes* energy to *break* a bond.

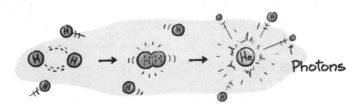

This simple mechanism is what lights up the entire universe. Because fusion is happening inside of countless stars, we don't have to live in a dark void. And gravity is what makes this possible, by pushing reluctant protons together until they fuse. But now comes the backlash.

The energy released by the fusion reactions rushes out, pushing everything outward and keeping gravity from further squeezing the protons together. Suddenly, we have two cosmic forces in an

---

* This fusion only happens if you have enough mass to generate the gravity needed to squeeze protons together. If you only have the mass of, say, Jupiter, then you just become a planet. If Jupiter had a hundred times more mass, fusion would start to happen at its core and it would become a red dwarf star.

epic tug-of-war: gravity fighting to squeeze everything together, and fusion releasing energy that pushes back on gravity. These two forces get locked into a solar stalemate that lasts billions of years.

The Long, Slow Burn

(4.9 billion years ago to 5 billion years in the future; Sun's age: 0.1 billion to 10 billion years old)

For the next ten billion years, the Sun is like an active war zone between two impressive forces: gravity and fusion. Gravity, the original actor in this drama, keeps squeezing and pushing together all the material in the star. But the energy being created by fusion pushes everything outward. The star burns and shines and lives on in this precarious balance for billions of years.

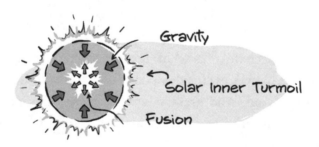

And this is where we are now. When you look up at the Sun (not directly, hopefully), you are watching a giant ball that is both exploding and collapsing at the same time. It's hard to grasp the scale of what's happening inside the Sun. Beyond the fusion core at the center, there are 350,000 miles of layers of hot, churning plasma. Photons being created by the core bounce around in these layers until, fifty thousand years later, their energy finally bursts free and out into space. Eight minutes later, some of them make their way here to give us sunshine.

The Sun has been burning this way for the last 4.9 billion years, and it will continue to do so for the next 5 billion years. The balance between gravity and fusion doesn't last forever, though. Silently, a clock has started counting down within the star.

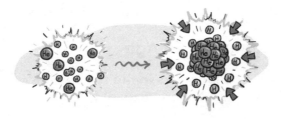

Gravity is weak but relentless. It will continue to pull on all the material inside the star forever. But fusion requires fuel (hydrogen) and produces waste (helium), which limits how long it can keep going. At first, the helium collects at the center of the star, slowly accumulating there and not bothering anybody. But eventually, it will start to change the star.

Helium is denser than hydrogen, and so the core becomes heavier, increasing the gravitational squeeze on the hydrogen that's now mostly outside the core. As a result, there are more fusion reactions in the outer layers, which makes the Sun hotter, brighter, and larger. The reactions grow slowly—every one hundred million years, the Sun gets 1 percent brighter. But it adds up. In four billion years, the

Sun will be 40 percent brighter than it is today, causing our oceans to boil.

As fusion burns even hotter, the Sun grows larger and larger. Fusion appears to be winning, but it's consuming its fuel faster and faster, and like a rock star on a bender, it will eventually crash and burn.

Getting Bigger in Old Age

(5 billion to 6.4 billion years in the future; Sun's age: 10 billion to 11.4 billion years old)

The battle between gravity and fusion lasts for billions of years, with fusion appearing to get the upper hand. Ten billion years after it started, fusion becomes so powerful that it actually reverses some of gravity's gains, pushing the outer hydrogen layers of the Sun back out.

At that time, about five billion years from now, the Sun will grow to two hundred times its current size, nearly enveloping Earth and all of the inner planets. Most of the Sun will be the fluffy outer layers of hydrogen, which will be cooler relative to the rest of the Sun. But by Earth standards it will be unbearably hot, and life anywhere in the inner solar system will be essentially impossible.

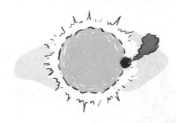

This dramatic display of fusion's power is its last hurrah. Having knocked gravity back on its heels, fusion overextends itself and starts to falter. But before it finally succumbs to gravity, it has one more trick up its sleeve.

One Last Fling

(6.4 billion to 6.5 billion years in the future; Sun's age: 11.4 billion to 11.5 billion years old)

At 11.4 billion years old (6.4 billion years from now), the Sun will have burned all the hydrogen in its core, exhausting the fuel that powered its battle with gravity. While fusion can continue to burn in the layers of hydrogen surrounding the core, it can no longer push against gravity's pressure inside the core.

Fusion isn't done yet, though. When gravity has compressed the helium core so much that the atoms get squeezed together, fusion starts to do the same thing it did with hydrogen. In a flash, it begins joining helium atoms together into heavier elements, mainly carbon. This is a literal flash, not a metaphorical one. When it's sparked, helium fusion releases as much light as the *entire galaxy*. Fortunately, this takes place inside the Sun, so the light doesn't burn humanity's colonies on the moons of Jupiter.

The carbon that the fusion reaction makes is concentrated at the core, making our Sun a triple-decker sandwich of carbon, then helium, then hydrogen. In larger stars, the cycle goes on to produce heavier elements.* But our Sun isn't massive enough to fuse carbon, so eventually the helium and hydrogen are exhausted and the Sun simply . . . fizzles out.

This period of helium fusion starts with a bang, but it doesn't last very long. While the Sun burned hydrogen for ten billion years, it will only burn helium for about one hundred million years.

Jupiter Goes Bonkers

(6.5 billion years in the future; Sun's age: 11.5 billion years old)

With all of its fuel exhausted, fusion sputters out. The outer shells of the Sun will drift out and form a nebula, the raw materials for future planetary formation. As fusion fades, gravity continues to work on the core, gathering the remaining elements into a very hot, dense blob known as a "white dwarf." This smaller star is about half of the original Sun's mass, but compressed into a ball about the size of Earth.

And this drives the outer planets, which survived the Sun's expansion, to a dangerous state. Because the Sun loses half of its mass, it doesn't pull on Jupiter and the outer planets as much. This makes

---

* For massive stars, the pressure gets so great at the core that carbon fuses into oxygen, which can fuse into neon, and so on. Each step is more rapid, but in the largest stars, it continues until it makes iron. Iron can't be naturally fused, because it would absorb energy rather than release it, and so this is the end of the road for fusion.

the gas giants relax their orbit to about twice the previous distance from the Sun. That sounds like a good move, given the Sun's previously fiery antics, but it makes those planets much more susceptible to gravitational tugs from nearby stars that pass by. In many scenarios, the orbits of Jupiter and Saturn become much more chaotic, ejecting the other remaining planets (Neptune and Uranus) from the solar system until only they remain. Eventually, only one remains, likely Jupiter, a solo gas giant orbiting the dead core of our Sun.

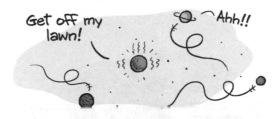

There's no fusion happening at this point, but the white dwarf still shines. Like a white-hot piece of metal pulled out of a forge, it glows from its own internal heat, and it will do so for a long time.

And now the Sun is stuck. The temperature isn't high enough to start fusion, and gravity isn't powerful enough to squeeze the atoms closer and upgrade the star further into a neutron star or a black hole.

The End (Trillions of Years from Now)

How long does a white dwarf shine? We don't actually know because we've never seen one dim. Physicists think that it may take *trillions* of years to cool down, eventually becoming a dark, dense mass known as a "black dwarf." But the universe just isn't old enough for any black dwarfs to exist right now.

This means that our Sun may exist as a white dwarf for a long time, maybe even trillions of years. It won't be as hot or as bright as it was in its youth, but it might be warm enough to sustain human life, once we abandon our temporary colonies on Jupiter and settle in closer to it. Perhaps as we sit around the embers of that white dwarf, humanity will tell stories of what life was like in our day, when the Sun burned and humans took it for granted. We'll reminisce about how it was a continual explosion, and it seemed like the sunny days would last forever.

Why Do We
Ask Questions?

O f course, we saved the best for last.

People have asked us a lot of very fascinating questions over the years. The range of topics varies a lot, from intricate and niche ("Why do photons get bent by gravity if they have no mass?") to profound ("Why does the universe exist at all?"). And in this book, we tried to give answers to the most frequently asked questions—the ones that tap into our shared curiosity about the universe and that seem to be most on top of people's minds.

But there is one frequently asked question that we haven't answered yet. In fact, it's probably the most frequently asked question we get. We saved it for last because we think it's the most important question we typically get about the universe. Are you ready? Here it is:

What does that even mean?

Okay, that's probably not the question you were expecting. It probably doesn't even feel like a complete question to you. Grammatically speaking, your high school English teacher would cringe. Still, it comes up a lot.

The interesting thing about "What does that even mean?" is that it's not the first question that people even *want* to ask us. Usually, we see this question added *after* their actual question. For example, people will sometimes write us and ask, "Hey, Daniel and Jorge, is the universe really fourteen billion years old? And what does that even mean?" Or, "Say, where does the energy to expand the universe come from? Can it really come from nothing? What does that even mean?"

In fact, we're guessing that "What does that even mean?" is not a question that most people were even *expecting* to ask. And yet there it usually is, added haphazardly to the end of whatever question they initially wanted us to answer.

At first glance, it may look like an afterthought or a random throwaway line. But actually, we think it's the most telling part of their question. That's because it reflects the true reason they were asking their question in the first place.

Here's what we think happens: People typically have an initial question that gets them curious. It might be about the age of the universe or about the nature of matter and energy in our cosmos. It could be something that they heard about in our podcast or read somewhere else. Whatever it was, it got the gears in their heads turning, and eventually it crystallized in the form of a specific question. But then as soon as the question left their mouth or their typing fingertips, a thought probably occurred to them: *What do I do with the answer if I get one?* And when they consider all the consequences that an answer might imply, a small voice inside of them whispers in their ear: *And what does that even mean?*

What does it mean that the universe is fourteen billion years old?

Or what does it mean that the universe is expanding out of nothing?

You see, it's not enough to know the answer to a question. The answer could be "yes" or "no" or "it comes from the Schwarzschild self-interaction of the vacuum Higgs fluctuations," but in the end the specifics don't matter. What ultimately matters is the *meaning* of the answers—the significance for how you live your life.

You might not think that the answer to the question "Where did the universe come from?" could change your life. But even if the answer doesn't affect the details of your life in a practical way, it can change something even more important: the *context* of your life. Fundamental answers can impact how you see yourself and how you relate to the wider universe. For example, learning that Earth is not the center of the cosmos made humanity realize that we are a small part of something larger, and that we are not on the universe's main stage. In the same way, discovering that the universe is filled with intelligent life—or that intelligence is extremely rare, or even that we are the only thinking beings in the universe—would deeply affect how we see ourselves and how unique we think we are.

It's that search for meaning and context that gives these questions their cosmic power. We want to not just *know* an answer; we want to understand it, because that understanding changes how we frame our existence. It can take us off the stage that we thought we were living on and reveal that we have been dancing on a completely different one.

The most tantalizing thing about the answers to scientific questions is that they are within our grasp. For each of the questions in this book, and for every science question you can imagine, there *is* an answer. It might be hidden, or far away, or at a scale too small for us to see right now, but the answers are *there*.

We may one day be able to answer all the questions in this book. But even then, we might have no choice but to append the same follow-up question that our listeners ask: *And what does that even mean?*

That's the one question we can't answer in this book. Why? Because the answer is different for each of us. We all get to define our own context and find our own significance in this universe. It's asking these questions that reveals who we are and why we search for meaning.

So what are *your* frequently asked questions?

Acknowledgments

Another frequently asked question we get is, "How do you find the time to write a book?" The answer is: with a little help from a lot of people!

We are grateful to friends and colleagues who reviewed early versions of the manuscript: Flip Tanedo, Kev Abazajian, Jasper Halekas, Robin Blume-Kahout, Nir Goldman, Leo Stein, Claus Kiefer, Aaron Barth, Paul Robertson, Steven White, Bob McNees, Steve Chesley, James Kasting, and Suelika Chial.

Special thanks to our editor, Courtney Young, for her continued faith and trust in us and her steady guidance; and to Seth Fishman for always finding the right place for our work. Thanks to the whole team at the Gernert Company, including Rebecca Gardner, Will Roberts, Ellen Goodson Coughtrey, Nora Gonzalez, and Jack Gernert, and to their international counterparts. Many thanks to everyone at Riverhead Books who contributed their time and talent to the making and release of this book, including Jacqueline Shost, Ashley Sutton, Kasey Feather, and May-Zhee Lim. We are also grateful to Georgina Laycock for planting the seed of the idea for this book (and the title!), and to the entire team at John Murray.

Jorge is grateful to his family, as always, for their continued support and encouragement.

Most of all, we are grateful to our readers, listeners, and fans, who have followed what we do over the years, and for their amazing questions.

Index

Page numbers in *italics* refer to illustrations.